PRODUCTION SOUND MIXER

*notes
& thoughts*

D1619794

Edgar Iacolenna

Production sound mixer
notes&thoughts
Edgar Iacolenna

Interpreter & translator
Chiara Salce

Additional translator
Federica Martelli

Proofreading
Christopher Farley

Graphic design
Vera Bianda

Front cover photograph
Oleg Magni

ISBN: 978-3-033-08992-1 (Paperback)
ISBN: 978-3-033-08993-8 (eBook)

www.edgariacolenna.com

CONTENTS

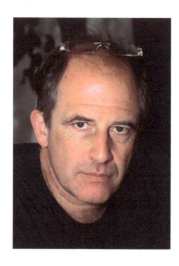

Michael Hoffman

Michael Hoffman is a director
and writer, known for
several movies including
The last station (2009),
Soapdish (1991), *Gambit* (2012),
One Fine Day (1996),
A Midsummer Night's Dream (1999),
The best of me (2014).

Photo by Kirill Pitersky

Foreword

My first film, an overlong student film called *Privilege*, which featured the debut performance of Hugh Grant (for which he has, quite rightly, never forgiven me), was also my first encounter with film sound.

Our sound man, Andrew Parker, was as green as the rest of us, but revealed, even then, the singular obsession I've come to depend on from his tribe. While the rest of us were capable of thinking only of the image on screen, here was an extraordinary creature who lived in a parallel universe of hums and creaks, sibilant whispers and unintelligible moans.

I remember the moment when it all came into sharp relief. I was directing a romantic encounter which was overladen with emotion. At the end of the take, I gave what was probably an insensitive note to a pair of lovers gazing too longingly into each other's eyes "The stares are a bit much, aren't they". I heard a voice from next room, it was Andrew, "Fucking nightmare" he said, "I'll put down some carpet".

It took me one moment to realise that he was talking about stairs, not stares, and the fact that heavy footsteps were ruining the dialogue.

AND THANK GOD FOR HIM!

And thank God for all the extraordinary sound recordists I've had the privilege of working with.

I am a director who cares very much about performance, but I can also, like virtually everyone on the set, forget that there is something going on beyond what the camera is recording. The sound man fights a lonely battle to convince a bunch of obsessives to realise that much of the core of the sensual experience that is cinema relies on good sound recording. The nuance that makes great performance can easily be lost if that solitary figure is not brave enough and clever enough to successfully fight his corner.

This book which celebrates these artists is long overdue.

The longer I've been in the business the more I've come to rely on my recordist.

One breakthrough moment came during prep for my film *Soapdish*. My dream was to create a film that called up the rhythm and music of an old-fashioned screwball comedy like *Bringing Up Baby*.

The script was absolutely laden with sharp, funny dialogue that wanted to be non-stop and rapid fire.

I had the cast, Sally Field, Kevin Kline, Robert Downey Jr., Whoopi Goldberg, et. al. to make it sing.

The first couple of recordists I spoke to, however, told me that I would have to be very careful about overlaps. They told me I should record everything as clean as possible and then create the overlaps in Post.

I felt strongly that to get the rhythm and energy I imagined, the actors needed to be able to overlap without worrying about it. Enter Petur Hliddal. Petur asked me to tell him what I wanted. I did.

He told me that with enough individual microphones and a certain amount of aggressive mixing, the actors could be given a great deal of freedom. He was very honest about the potential compromises. He told me there would be moments that would demand clever sound editing and that in other moments the sound might not be as pristine as some recordists might demand, but he wanted to help me realise my vision. He gave me a confidence that I communicated to the cast and together we created a film that has been called one of the most successful farces ever put on film.

I went on to work with Petur on *One Fine Day* and *A Midsummer Night's Dream* in Italy.

On that project, he was the only person on the crew that I brought from America. That is testament to how strongly I felt about him a collaborator, how central he had become to my creative process.

Over time, the way I work with my recordist has changed, I now discuss the overall sound scape of the film with the recordist as early as possible. That means, in addition to very good dialogue, I end up getting all kinds of gifts in terms of ambient effects. I have also learned to trust the recordist as my second set of ears, not only in terms of audibility but also in terms of the subtlety of performance. It is the sound recordist more than any other crew member that I go to when I have questions about the quality of performance, because their focus is fully on what no one else is caring about.

I have made films in Germany, Italy and Quebec. While the language of the sets was largely English, most of the crew members were not native speakers. In the beginning, I felt that it was important to have a native English speaker recording dialogue. Last time out however, I worked with Adriano Di Lorenzo, an Italian speaker, and came to find in our constant conversation about the nuance of words and phrases, that I was forced to see new shades of meaning in the text. What Adriano heard and the way he heard it supplied me with a fresh set of ears on certain things I had taken for granted; it was a rich and rewarding partnership.

I have long been impressed by the grace and subtlety of the recordist. No one else on the crew has to deal with the impatience and frustration from other crew members simply for doing their job: an actor will ask for another take, the cinematographer will not hesitate to stop things if the sun comes out, but God forbid the sound recordist says we might want to go again. It takes real belief and bravery to stand up to the time pressure and the shrugs and sighs of the producers and ADs, to stand up and say: "We don't have it".
But in the editing room it is the brave sound recordist whose praises are sung.
Thank you, all of you, Louie, Adriano, Simon, John Casale, Petur, Andrew, Patrick, all of you, for being smarter than the rest of us and protecting our sound, this critical aspect of the cinematic art that can so easily be abandoned and abused by the rest to us.

8

Prologue

When I was little, like many other children, I was afraid of the dark and especially of burglars.

I couldn't fall asleep straight away and so I triggered a defence mechanism of listening.

I listened to everything around me, including the silence, because I had to be ready to react at the first suspicious sound. Fear caused me to be very focused on listening and to associate certain sounds or noises with certain things, without really knowing whether it was something real or just my imagination drawing the wrong conclusion.

I used to solve the problem with a surge of courage, running to take refuge in my brother's or my parents' bed. But, even so, I vividly remember the moments that preceded this step, and, upon reflection, I consider this my first introduction to active, critical listening, albeit imaginary.

As my father is a musician, I went to his concerts from a very young age.

I was certainly not ready to "understand" all that music can convey, but what came to my ears, in one way or another, began to shape and define certain sensations for me, making my hearing sensitive to every nuance.

A large orchestra in a concert hall or an organ concert in a cathedral will certainly have a strong impact on a child; even if they do not grasp or feel a sense of musical emotion, the sound can definitely convey a lot in itself. Such sensations unconsciously build our sensitivity to listening.

Only now, many years later, do I realise the part these childhood experiences played in the way my ears have developed and how I approach listening today.

Of course, this is nothing extraordinary or exceptional, but simply a reflection on the awareness of how certain experiences can mark and direct our path. Looking back, we discover that so many of those moments in our lives that we did not value, except as mere memories, were actually the building blocks of our development, as people and as professionals. Other memories come to mind.

A simple and wonderful game that all children play; hide and seek. Yet, even in this, sound and listening were involved.

During the count, my friends would hide and, while I was counting with my eyes closed, my ears became my most important sense.

Even while counting, you could sense the slightest movements from the other children as they headed for their hiding places in all directions, making it easier to locate them.

A trivial thing? Maybe, as it seems so obvious to us and we take it for granted, but, in reality, it is an example of how our hearing provides us with thousands of pieces of information in every single moment. We decode this information automatically without realising it, so it seems like a natural, obvious process that we pay no attention to, when, in fact, it deserves special care and attention.

When I was ten or eleven, I started listening to various genres of music.

Fortunately, there was no shortage of electronic devices at home, as well as various types of headphones and other interesting gadgets. I didn't pay much attention to them, but they were always at hand.

Of all of them, I remember a wonderful Walkman that began to be part of my daily life and that I could not do without.

It had a button called XBS (Extra Bass System) which increased the bass frequencies considerably. A new world opened up to me. It made me so happy. The music had such crazy depth and I especially enjoyed listening to certain kinds of music.

However, I noticed that each of the different headphones I had at home "sounded" different, some were able to cope with this heavy sound, others were not and struggled badly.

Naturally, I chose the headphones that went best with that Walkman, and I used these faithfully for a long time.

Looking back, I'd made a choice based on taste.

Obviously, when I was eleven, I liked that kind of heavy, booming, brooding and totally unbalanced sound.

A few years later, though, I realised that I wasn't interested in that kind of sound anymore, I found it strange and unnatural, and wondered how I could have liked it so much.

People change over time. Our personal taste gradually develops, transforms, evolves and can be drastically different from what it was before.

A song or record that we didn't like at all before may seem amazing two years later, so we re-evaluate everything.

After spending my youth in music, Walkmans, films and professional football, I am now a Production Sound Mixer and Boom Operator.

Fortunately, I still have that spark, that passion that makes me curious about sound, eager to learn about different ways

10

of working, of thinking, eager to hear the thoughts of other professionals about this fantastic profession, namely being a production sound mixer and boom operator.

My personal way of seeing this career is linked to a fundamental concept that I have developed over the years. The work is based on a simple action that we can sum up in one verb: recording.

For me, this word means being faithful, putting yourself at the service of what you are recording, trying to capture every nuance with the intention of doing your utmost to protect what you are hearing, to do it justice, knowing that that performance, that sound, that speech will be immortal and, therefore, remembering that our work can affect the quality of the recording itself and the way it will be preserved for posterity.

It is clear that behind this single word, recording, there is a whole world; a world made up of technical and taste choices, of "compromises", culture and memories; a niche world that is obscure to those who do not deal with sound.

George Martin, the Beatles' famous avant-garde record producer, wrote a book, the title of which, first of all, left a deep impression on me: *All you need is ears.*

I believe this should be the starting point for every sound engineer: their own ears.

The ears are the primary tool a sound engineer has at his or her disposal, certainly the most important tool, to be educated, developed and preserved.

Needless to say, it takes many years for our ears to mature and to develop a subjective, unique and special connection with sound.

Personally, in recent years I have felt a change in myself, a different approach to sound, a continuous and deeper search, which I imagine will carry on for a long time to come. The ultimate goal is to build our own personal taste, perhaps taking a certain type of sound, a sound we like, as a model and trying to pursue it in our work as sound engineers.

I like to describe this process as a rediscovery of our own sound consciousness. How do we go about developing our sound consciousness? First of all, by improving our critical listening and quality assessment skills. We start to ask questions about everything that comes to our ears and try to give answers, but by looking in detail and not just on the surface. Why do we like that particular sound? Why don't we like it? Why do I like the sound of that microphone more than another? Why do I consider the quality of that sound to be poor? It may seem trivial to ask these questions, but this is not the case. It is not easy to understand, analyse and describe why our ears and brain cause us to prefer and opt for one type of sound over another, but trying to analyse these aspects and understand the reason for our choices is the first step towards building a personal taste.

All of this, I believe, applies to any specific area within the fantastic world of "sound".

It is precisely this continuous search, this "hunger" for sound, made up of moments of exaltation and moments of profound despair, that has led me to write this book.

What is this book?

It represents the "journey", in a figurative sense, that I wanted to undertake in order to discover the various ways of understanding this profession; the various ways of approaching the profession of production sound mixer and, therefore, the recording of sound on set.

I had the opportunity to interview seven international sound engineers, chosen by me on the basis of a simple criterion: I have appreciated their work and their career, and I was therefore interested in being able to talk about "sound" with them and explore some specific topics.

The "stages" of this journey took me from one part of the world to another: from the United States to India, from Australia to France, from Italy to the United Kingdom and, finally, to Switzerland. The fact of being able to talk to sound engineers of different nationalities has inspired me even more, since my aim with this book is precisely that: of revealing, through the interviews, some different ways of working, as well as shared aspects. These points of reflection can professionally help both a young novice who is preparing to enter this career and a "seasoned" professional who can, perhaps, consider new possibilities and new methodologies.

I deliberately asked some identical questions to each sound engineer in order to highlight the similarities and uniqueness of their ways of working and thinking.

Travelling is a way to "open" your eyes, to discover new possibilities and, sometimes, to rediscover yourself.

I therefore believe that, thanks to this "journey" of interviews with these leading professionals, readers will be able to develop or shape their own ideal of a production sound mixer, perhaps starting out in this profession on the right foot or finding new inspiration to continue it in the best possible way.

Summing up, I can say that I have tried to produce the book I dreamed of reading when I was a student on the sound course at the Centro Sperimentale di Cinematografia in Rome (cinema school), so I hope that a book like this will be of help to many students who will have a future in the world of cinema sound.

I would really like to thank, from the bottom of my heart, all the sound mixers who helped me with this project, giving me their time, according to their availability, even though they were often clearly busy on set.

To them I extend my gratitude and esteem.

Tod Maitland
United States

Guntis Sics
Australia

Chris Munro
UK

Pascal Armant
France

Stefano Campus
Italy

Nakul Kamte
India

Patrick Becker
Switzerland

Tod Maitland
United States

18

Tod Maitland

Tod Maitland is a four-time Academy Award nominated, BAFTA winning sound mixer with over 100 feature film credits. His films range in time and diversity from *Tootsie* to *JFK*, *Seabiscuit*, *I am Legend*, *Across the Universe*, *The Irishman*, *Joker* and *West Side Story*. Always at the forefront of technology: Tod was the first mixer to use Dolby SR on location, is currently ushering in the new age of wireless systems, and has become a leader in the art of capturing sound for modern movie musicals. "From when I first worked on *The Doors* with Oliver Stone, I was hooked, it doesn't get much better than that". For Tod, *The Doors* began a 30-year journey to make production sound better on musicals: better sounding, better efficiency and better technology. He has filmed 11 musicals to date.

Beyond production sound, Tod has collaborated with Wylie Stateman and Lon Bender to create "The Hollywood Edge Sound Libraries", the largest in the world and is also the head of Sound at NYU Tisch Film School. Tod splits his time with his family between NYC and the country where his hobbies included inventing, writing and building things.

The *West Side Story* sound cart - 2019

Before becoming a production sound mixer, you worked a lot as boom operator. Has this experience helped you? Do you believe that in order to become a good production sound mixer it is important to have a practical experience as boom operator?

As a boom operator you learn microphones.
To me, 80 percent of the art and ability to capture good dialogue, or any sound for that matter, is microphone placement. If you don't have the microphone in the right place, it doesn't matter what equipment you have, it doesn't matter how good you are as a mixer, if that microphone is not in the right place, it won't sound good.
As a boom operator, the first thing you learn is what a boom microphone can capture, and what it can't capture, which is just as important. For example, when you have a microphone one foot away from an actor and it's an acoustically friendly environment, the sound is going to be wonderful. Conversely, if you have a microphone one foot away from an actor and you're on the city streets of New York, you may have a real problem. Then again, if you're outside in the quiet countryside, you can have the microphone 10 feet away from an actor and it can sound great.
Being a boom operator taught me, as a mixer, what's possible with the boom mic. Now, as a mixer, when I watch the rehearsal of a scene (aka a blocking rehearsal, which btw, is vital to me analysing how to capture the sound in that scene) I can quickly assess and figure out how to mic the scene and any alternative miking if necessary. Also, my boom operator watches the blocking, so together if they are having problems with shadows or cues or the angle of the microphone, or depth of field... I can help them. Things you learn from booming are crazy, for example, if an actor talks directly into a window, it's tricky, especially when their face is close to the glass. Through trial and error, I learned that instead of having the microphone straight up and down towards the actor, I would actually aim the mic into the glass and pick up their voice off at the angle of reflection of the glass, and it sounded great. There are other tricks that you learn as a boom operator; many if it were a one boom scene, you were actually doing the mix with the mic.

There was a sound mixer by the name of Nat Boxer who was the boom person but he was technically the sound mixer, he got the credit as a sound mixer. Nat did *Apocalypse Now* and took home the Academy Award for it – as the boom operator. Nat felt by being the boom operator he could do a better mix with that single microphone than he could do sitting at a mixing board, and I believe that's mostly true back when mixers recorded to a mono track when you filmed with one camera and generally one boom, so you were only capturing sound for one perspective. Now we film with multiple cameras and capture sound for a wide shot at the same time as a close up. Unfortunately, that old school way of working, which for sound is much better, is over. Now you have to capture sound for two or more perspectives which forces a mixer to use wirelesses in situations where they wouldn't otherwise have to. Generally, most mixers wire everyone now just in case.

Having done this for so long, I'm in a lucky position, I can push back and say, "That's not the sound that we're going after, and if you change this or that we can't make it work for sound". It's a forced collaboration. A lot of sound mixers will get railroaded by production, saying "We don't have time for the boom to work it out", or "Just wire everyone". I feel very bad for a lot of production mixers, it's hard to stand up and you really do need to stand up for yourself. On the other side of the coin, Camera can hold up production whenever they need, but for sound, it's not the same.
To cut a long story short, I do believe that being a boom operator was instrumental in understanding what microphones can capture and in what situation, I learned a lot about hiding microphones also.

Looking over your filmography, it's impressive how many world-renowned films you have worked on. How does one reach such levels, what characteristics should a production sound mixer have to reach and maintain such levels?

First of all, I was very lucky because I was born into the business.
My father was a sound mixer, so I immediately started working on bigger films instead of having to work my way up through student and independent films, understanding how large films work was a big early advantage. But, you know,

it's that old rule – first you get your foot in the door, then you need to prove yourself. One of the things my father taught me was – half being a good sound mixer is what we know technically, how to get sound and all that, the other half is how we play the game. What he meant by how we play the game is how we interact with all of the other players we meet on a film set. As mixers, we interact with directors, producers, actors, prop people, electricians, grips, everyone. You have to know how to communicate with all of them, and they're all different. It's how you stand out from the crowd.

One thing sound mixers do is "cry wolf", meaning, go to a director for every problem. That devalues the moments when you really need help. It's very important to know when you need to go to a director and say: "This isn't working, we need to figure it out". I think those lessons are as valuable as all the other lessons. Working with bigger actors is part of this also. It's easy to be intimidated but you need to be able to get the sound you want from them, even if it means adjusting their lav multiple times between takes. We all get a little intimidated, I mean, even I do at certain times, but I always try to push that aside and remember that I'm there for a reason, and that reason is to capture the best sound for this film. Bigger films have bigger budgets, which have bigger responsibilities and more pressure, but they also have more resources, so on a big film you will have a lot of resources, you learn to use them like other departments do.

23

Another thing I do is go on all the location scouts. I think that's another part of filmmaking where sound mixers get pushed aside. I believe these early moments are crucial, not only because now I get to see the set and listen to the what the director is thinking about for the scene and what the cinematographer is thinking about, lighting wise and all of that, but it's about developing a relationship with the director and producers and department heads before you start filming. Location scouts are of course also about finding sound problems, there are a lot of inherent noises we deal with; air conditioners, refrigerators and all of that is important.
At this early stage in the film, everyone is not so anxious, you have time to talk, you have time to develop a relationship, and once you've developed a relationship, when you go and talk to them later, when you have issues, you have more credibility, you have more of a connection with them.

I believe that's one of my attributes, is my ability to communicate with fellow filmmakers, make them feel good. For directors, let them know I'm there with the same goal as them – great sound.

Besides being very good at your job, how important is it to cultivate personal relationships and know how to navigate this industry?

I think I touched upon that our business is a very social business, you're going to run across many people you like and some people you don't like, but everybody is there for one purpose.
I always try to develop relationships, certainly with all the people I know I'm going to need their help for me do my job well, this can be everybody, for instance the prop person, if the actor is working with a grocery bag and it's making a lot of noise then I need help with that grocery bag, or the wardrobe person because a piece of clothing is making noise, or the cinematographer or the camera operator. We all interface.
Every one of us interacts with each other, which is what makes the film business really very special.

We all have very specific jobs, but we are all very much a part of a web, and without all of the connectivity of that web, like a spider web, it doesn't work as a whole.
I find it vitally important to develop those relationships. Some mixers are very good with technology, they can talk about a transmitter or a diode or a piece of cable very well and they know exactly how to fix it, but they don't know how to communicate with people, and if you don't know how to communicate with people, it's a real detriment to our business.
I believe, as I said before, that 50 percent of what we do is how we present ourselves and how we integrate with everyone around us. So, knowing how to work technology is of course very important, but the interfacing, the being part of something, part of a whole as a group, is what I've always loved about the business.
You share in experiences, whether it's one of tremendously difficulty or fun, most of the time they are rewarding, you've shared an experience with people.
I think some of the most difficult films I've worked on have shared the experience of going to war together – obviously you can't compare the two but maybe you were in zero-degree locations weeks on end, but you survived it, and now

you have this camaraderie, you have this common bond.
I feel that relationships are incredibly important in this business.

When you record sound for a movie, what is your objective? What do you try to provide to the post-production team to ensure a scene is complete (from the point of view of the sound recorded on set)?

When I approach a film, and every film is different, I read the script. But my approach is almost always the same. I'll give post-production as much as I can possibly give them.
What that means, particularly these days, is we put a lot of mics out and wire everyone most of the time. I was one of the last mixers to surrender to this "wire everyone" philosophy. If I knew we could boom a scene, for me there was no reason to put wirelesses on them.
It would drive me crazy that they would always want them wired, post-production got used to having sound mixers wire people because maybe production wouldn't take the time or work out boom shadows, or they're boom op wasn't good, or whatever. The problem now with directors is you may be shooting a scene where you could boom the scene and then all of a sudden, the director says: "Oh, let's pull the camera back and let's do a wide shot of this". And now you have no wireless on the actors, so you have to stop everybody to wire the actors and that's a big problem. Anything that consumes time is a problem.
Unless it's something where I know for sure that I'm not going to have to wire the actors, I just end up wiring them all the time, which post-production really likes. I've come to accept this and even embrace it. Life is about options. The more options that you have, the better off you are.
In fact, when I teach at NYU, I teach my students to give themselves more options in Post. Which means wiring everyone and booming them also.
When I work, we'll always boom on and off camera actors, but we also keep ambient mics going just to capture the general atmosphere of the scene. This is particularly helpful when scenes are both wireless and boom. The ambient mic on the set allows me to hear the bounce of the room. If you have an ambient mic maybe 15 feet away, you can hear the bounce of the room and by mixing it in you can make wireless mics sound like a boom.

25

Having post-ambient mics is a gift, for instance, on *West Side Story* we had a lot of period cars, we had a lot of period props, we had kids running in the streets with period shoes…
We always keep a couple of extra microphones with wireless transmitters on them so that we could just put out if we see something that sounds interesting while we're filming; my team was always great at seeing opportunities.

If we see a car going by over there, and background kids are playing some game over here, we'll throw a microphone out on it just out of frame and put it on a separate track.

I'll write on metadata what it is so that they know in post-production and so then I'm just giving them more options.

To me, it is so much about options.

Like I said, life is options, and I believe that the more that I can give Post in options, the better off we all are.

There are times that we'll do recordings after we filmed the shot.

For instance, on the film *Joker*, a good chunk of it took place on a subway car, to capture those sounds I took an ambisonic microphone and spent days on subway cars. I would place the microphone in between the cars, I would have it in the car and record with the window open, window closed, empty, full, just to give Post many different options.

When I record sound effects, I always try to record sound effects with one microphone close and one microphone further away, giving Post two perspectives. This allows them to play with perspective in their mix, it can sound either close or sound further away. Those are recordings that are difficult to recreate in Post. Always try to record as much as you can in production.

When I do a musical, if we're doing all lip sync with speakers playing on the set, after the scene is done, I'll give all the actors earpieces and I'll have them recreate the entire dance scene without singing, without any vocalization, allowing us to get their footsteps and their breathing and their movement and all of that, in essence, it gives you an in-sync Foley track.

Something else I do is I'll also get an actor's voice in my head, like Robert DeNiro.

If I know how he sounds, then I can record him in different locations i.e.: a bathroom, a field, in a car… locations that are going to affect his voice quality and require using different microphones, different techniques, different styles, EQ, but I need to make his voice sound the same: we film a movie in 4 months but we watch it in two hours, he has to sound the same. That's one of the elements of being a mixer. I use equalization a lot also. To make actors sound the same you need EQ. Some microphones require different equalization than others. Other things that effect EQ are ambiences, acoustics, wireless mics under clothes sound different then wireless out in the open, so you have to EQ each to sound the same.

The biggest thing is using your ears, you have to use your ears.

You have to listen to what sounds good, and what doesn't sound good. Also, listen back after you've recorded something. When you're filming, your focus is on many things, when you're just listening to it you can hear things you didn't hear previously because you were focused on the moment.

Having grown up in this business, my father would always say that sound is an art, the microphones are the paint and the recorder is the canvas, that's the way I try to approach it also.

Sound has definitely become more technical, I am the least technical mixer of the sound mixers out there, I know what the equipment is supposed to do, but if you ask me to take it apart and put it back together, not a chance. On the contrary to that, I always keep up to date with technology because I need the advancement for the projects I do.

When recording, what makes you happy and satisfied and, on the contrary, what causes you to be disappointed or dissatisfied?

My mood can go up and down depending upon how good the recording is.

If I know that I could have done better, I'm always disappointed and that will stay with me for an hour or so or even longer if it's really bad.

But if I know that I was able to capture those moments, get those microphones in the right place, and I love the sound of their voice and I've been able to do a really good mix where I feel like we really nailed it, it's a wonderful feeling.

Almost every day you get the good and you get the not as good. Some films you work extra hard for no reward but no matter what, I can't do a bad job. Even if I'm working on a film that I hate and with people that I hate, I still can't do a bad job, it's just in my personality to do it well. I think growing up in this business was a big part of that.

When I started, this business really took care of everyone, I mean, we were a family and we were all paid very well and the studios were owned by families, it was not a corporate business like it is now. Now the business is very different, it's very corporate, the business doesn't care about the individuals like it used to and a lot of the younger crew members reflect that in their work.

I see the kids coming in and they're sitting on their phones in-between shots instead of watching the actors or watching the film making process, they'll do their job and then they'll just go on their phone.

I was talking about this recently with someone on set who I kind of grew up in the business with, we vets seem to have a different philosophy.

Ultimately you have a responsibility to the film, but younger crew members may feel that the film companies don't take care of them that well, so they're not going to sacrifice themselves in the same way that we sacrificed ourselves. I understand that. The business has changed but I'll still do almost anything for the project.

I imagine that throughout your career, your sound cart has changed many times, as well as your equipment. Can you give us an overview of your professional tools/ equipment, what they looked like in the past, how they are today, and if you have a piece of equipment that has always followed you on set in all your movies.

In the past everything was about the recorder, there were not many wirelesses. In those days they only filmed with one camera, so the boom and the recorder were really the primary pieces of equipment on a film.

Then as films got more complex, sound mixing boards came in that had eight channels of mixing, but you're still mixing down to the two tracks on a stereo Nagra.

In other words, if you didn't open up the right fader at the right moment, that sound was not going onto that tape. That's back when we really were mixers.

Just before I got into the union, everything was mixed on a mono Nagra, everything went to one track!

Now all our sound is on isolated tracks, no matter how the sound is mixed on set, it's going to be recorded.

My transitions were from a Nagra to a multitrack mixer to the early digital recorders.

I refused to use the DAT recorders because I felt that their quality was very poor, I would use DAT's for sound effects only, but I would never use them for a movie.

When disks came in, I switched over to them, mini digital disks in 2002.

Actually, before that, at the end of the eighties when I was doing Oliver Stone films like *The Doors*, *JFK* and *Talk Radio*, I was using Dolby SR.

Dolby SR made analog tape sound beautiful.

It's still the most beautiful sound out there, it's so warm and lovely, Dolby SR eliminated any tape hiss, there was no tape sound, but you had all the warmth and beauty of an analog tape. I loved that.

But it was a monster to haul around, it was one-hundred-pound machine that I would drag with me everywhere. Even on *Born on the Fourth of July* I dragged it through the desert with me. On top of that, I would have 2 stereo Nagras side by side and link them together so that I could have four tracks.

Then digital came in, I started with the Fostex minidisc, then the eight-track digital machine came in which just made everything easier and sound better.

Wirelesses were still not as good, but they were getting better. And then cut to 2 years ago when Steven Spielberg called me to do *West Side Story*.

Knowing the *West Side Story* script where 20 actors were talking in every scene, singing, dancing, fighting, all of that, I realized that I needed to change my entire sound cart.

For 25 years, I used a Cooper mixing board, which I really liked, it sounded great, I used a Zaxcom Deva for my eight-track recorder and Audio Limited wirelesses, they all sounded very good, but I needed four times the number of channels and equipment to do *West Side Story*, so from the ground up I rebuilt an entire cart.

I literally spent four months designing and building the new sound cart.

Since everything is wireless now, and we live in a world of wireless: the boom is wireless, lavs, my music speakers are on wireless, every monitoring system is wireless, so I realized to build this new sound cart, I needed to start with wireless.

I worked with Gotham Sound in New York, they are the sound company that we all use.

Peter Schneider who heads up Gotham Sound, called in all of the big wireless companies, every one of them came in and they did three-hour demos for us.

30

Testing various lavalier microphones with a Sennheiser
MKH 416 shotgun mic overhead for comparison

The company that I ended up going with was Shure. They were ahead of everybody: their transmitters are tiny, they're lightweight, there's no antenna on them, they're waterproof, they're so easy to work with and with their wireless workbench system it's amazing, I can literally push a button on my cart and it'll find twenty four free channels in New York, which is not easy, and with one push of a button, it will deploy them to all twenty four units.

Now all I have to do is start wiring the actors and I don't have to think about all the other wireless aspects and problems that I used to have.

This system also figures out my IFB system frequencies for headphones, for the director's headphones, my boom op headphones, for my wireless speakers...

Now I have everything in one place, everything is fully digital with endless possibilities, everything is also now Dante connected, so there are no more XLR cables going back and forth. It's all very neat, clean and complicated because now you have five new operating systems to learn.

I had to learn a digital mixer, I have an Alan and Heath digital mixer, I had to learn the Shure system, I had to learn new recorders... Not necessarily what I had planned on doing at this point in my life, but I have to say, I'm so happy that I did. It makes my life so much easier knowing that I can trust my equipment and I don't have to go through those wireless problems; they were always the biggest ones.

It doesn't change the aspect of filmmaking in any way. It just makes my life easier.

If there's one thing that I've kept my entire time, it would be my Sennheiser MKH 416.

It is my go-to microphone. I try to match all my microphones to it.

One thing I've learned also is that when you're using all these different wireless systems, every lavalier mic sounds a little bit different, and they sound different on different actors. Now I do a test before every film: I'll set up a horizontal bar that has six different lavalier mics on it - a Sanken, a Sony, a Shure... and I'll have a 416 over the actor's head. I'll record the actors on all mics. If it's a musical, I'll have them sing and talk, if it's just a speaking film, I'll have them talk and then later I'll go back and I'll listen to see which lavalier matches the 416 the best, and then I'll choose that lavalier for that actor.

So, whenever we wire that actor, that lavalier goes to that actor.

The 416 is definitely the only piece of equipment I've maintained throughout all of it.

I held on to my Cooper mixer for as long as I could, just because I love it but time goes on.

Edgar

It's incredible, I have seen the picture of this test that you've made with the lavaliers and 416 on top, and it is impressive, this search for the right sound is really incredible.

Tod

Yeah, it was by accident that I realized that.

I used to wire all my actors with one microphone, a Sanken microphone. Then I got a couple of these Shure microphones. I'd wired one actress with a Sanken and then the next day she left the Sanken in her trailer or something, I forgot, so I just happened to have a Shure and I put it on her and I was like: "Wow, that sounds entirely different", it was a whole different sound and it sounded much better for that particular actress. That's when I realized and I started doing these tests.

It went further, I realized that sometimes different lavs sound different when an actor's singing from when they're talking. That's became a constant thing that we do now.

I'm actually doing a musical this spring with Will Ferrell and Ryan Reynolds (a comedy for a change, thank goodness - everything I do is so serious!).

Anyway, I'll be doing the exact same thing.

I do these tests prior to the actors doing their vocal pre-records on a musical. Then I'll send those microphones to the vocal recording studio where they're doing their recordings and they will record 3 vocal tracks: my lavalier microphones, my 416 and their big fat studio microphone.

In this way, Post can transition from whatever microphone I was using in the scene to that same microphone in the studio, and then during the first line of music, they can fade into the big fat microphone so it's not a big jump from a wireless to the big microphone.

How many people make up your standard sound team? Which characteristics should a good boom operator and sound utility have? You managed to find professionals who have worked with you on many films, it's great when such relationships are born and last throughout the years.

I normally just have two people with me unless it's a musical, but normally just a boom person and utility person. It's a variation of Mike Scott, Jerry Yuen and Terrence McCormack Maitland, my nephew and hopefully heir to the throne. Back in the day, the boom person only focused on the set. They lived on the set and communicated with me everything that's going on.

I always watch all the rehearsals and talk about how to do the scene, but once I go back to the sound cart, the boom person was my eyes and ears, always thinking about what's the best way to boom the scene or if we need to hide a microphone, and if there're problems going on lighting wise…

The normal traits of a good boom operator are – they have a good eye, they have good perspective, they can get the microphone in the right place.

The other is knowing how to work the angle of a microphone, if actors tilt their heads down, if they turn, those are all standard things. Plus, they need to memorize the dialogue so they know when to go from one person to another, and they need to know the dynamics of a microphone, how far can they be from the actor before it doesn't work. They also need to be good communicators.

But now that we shoot multiple cameras and wirelesses are more prevalent than they ever have been, that person also needs to be knowledgeable in wirelesses and other technologies especially on a musical where there is a lot of other technical stuff going on.

My utility also has to be dexterous, a good 2nd boom, a cable runner, technological beyond me, able to seal up a window, turn the refrigerators off and mix in addition being a computer technician. Of course, they still have to be on top of eliminating sound problems on set, such as refrigerators, air conditioning, or a woman walking in heels, which in that case they need to put plugs on the bottom of their heels… Sometimes the boom operator and the utility person's jobs cross over, actually now more than they ever did in the past.

33

Tod Maitland, Terence McCormack Maitland, Jerry Yuen on *Tick Tick Boom*
directed by Lin-Manuel Miranda - 2020

I always want to work with the same crew. I'm very lucky, I really like my crew, I'm also lucky because I get to pick and choose good projects to work on.

At the moment I'm off and my crew is working on other shows. I usually know what film is up next about 4-12 months ahead of time, my crew tells whoever they're working with that moment the timing. That's the deal they always work out. Other sound mixers are very happy to have them whenever they can because they're very good, any mixer would rather have them for three months than not have them at all.

Most of all, you need to have fun with your people, you need to be able to go out for a drink after work, you need to be able to laugh, you need to have a relationship with them, it's a very important relationship because you spend so much time together.

The amount of time we spend together is more time than I spend with my family, so it's super important. If I had a bad relationship with someone, if you didn't want to see them on a daily basis that wouldn't be fun.

There have been films where I've only been able to bring one person with me. That's really difficult for me. We're a tight unit.

It's difficult to find good people, when you find them, you want to hold onto them. I'm lucky, I do get to have projects that people want to work on. Nothing wrong with that.

In an interview I have read that you also place the wireless microphones on the actors, and that you enjoy doing it. Does it happen often? Today, with the use of several cameras at the same time, it would be very difficult to secure the sound of certain scenes without wireless microphones. How was it in the 80's?

As a mixer, my focus never deviated from microphone placement being the single most important element of sound. So, yes, I want to be the one who places the lav on the actors. As I said, I truly believe that microphone placement is 80 percent of what we do, if I know where that mic is, I know how to mix for it. Ultimately, if you get a lavalier in the right place with no clothing noise, no wind noise, it makes life so much easier. The best is to have a boom in the right place at the right angle with a good boom operator who can follow the actors and who knows the dialogue, then your job as a mixer is so much easier.

The follow cart made by Jerry Yuen

On the converse, if a microphone is just not in the right place, the wire is screwing up, your job as a mixer becomes 10 times more difficult.

Back in the 80s, you only had one camera, if it was a very wide lens, you would talk to the director and the director would say: "I'm only going to use this as an establishing shot for the first five seconds, so don't worry about getting very close sound, I'll be in for the coverage for when there's the dialogue". If they said they were planning to use the dialogue from the Master shot, you would wire the actors, hide mics, whatever you needed to do. I actually like to hide microphones a lot.

It was always an art to hide them, a game to try to predict how well they'd work. They don't call them plant mics for nothing. But we always tried to boom everything.

I used to use wireless in the 80s, we just didn't have to use it as often because, again, you only had one camera and most directors wouldn't play a whole scene on a wide shot, it was a different way of filmmaking than it is today, and cameras just didn't move as much as today. Today a camera is another character in the story, we use technocranes so often, you can have a close-up transition into a super wide shot in the same take. Just the way it is.

Nowadays, in 2021, does it still make sense with work with cable? Yes? No? Why? Can you give me some examples?

No. Period. Wirelesses are so good and you don't have the time that you had before.

You used to have more time on a film set, now if you're spending time running cables around, you're not using that time efficiently to capture better sound.

I was one of the people who was last to go wireless for booms and everything, but now I would never go back.

The difference is you that you can't tell the difference anymore. Wireless is that good, the ease and ability are amazing. On my sound cart I always have three booms ready to go with wireless set on them. We can just turn them on and go. In the cable days, if I needed another boom, we'd have to run cable to where the other boom was. This way now the second boom operator can just grab a pole and go. The efficiency far outweighs the very, very, very minuscule difference in sound quality.

It's all about time, and the more time that you can save production, the better you are, and the better they look at you to. I pick and choose my battles, and this one is not worth fighting. It comes down to efficiency and now we have really good wirelesses.

Honestly, I could put a cabled boom and wireless boom side by side and I bet I couldn't tell the difference.

Of course, in my heart of hearts, I still wish I was using Dolby SR with cabled mics, but…

Have you ever experienced unexpected circumstances on set which you were unable to solve? How important is it to plan, organize one's work and foresee as many difficulties as possible beforehand? After reading a script, how do you proceed?

I think you're always going to come across unexpected problems and obviously you want to try to plan for as much as you can. On location scouts, I'll try to figure out what the problems are potential issues in different locations, but there are always going to be variables.

Part of the reason we're good at what we do is because we can think on our feet and come up with solutions for those things unforeseen.

I have always said that if I was ever trapped on a desert island, I would like to be trapped with a film crew because they are very resourceful. They will figure out how to do things that have never been done before, every time we do a shot, it's different from the last shot.

I think one of the good attributes of being a filmmaker is the ability to pivot, to adjust, to work with what you're given to create something, and it's not always going to be optimal, many times it's not even close but you still have to make something out of it, and you have to come up with an answer. I like that part, I'm good at that part, I like problem solving, I like making things work that weren't necessarily made to work in a situation.

As with a script, you can only get so much information from a script, you read the script, sometimes it's basic, pretty laid out. All of course depending upon the director's vision, it could go in so many different directions.

The relationship with other departments is crucial. If everyone worked not only in the interest of their own department but also helping each other, everything would be easier and more fun. Have you ever found yourself struggling with other departments?

Very very few times have I struggled with other departments. I did work on one film where I had a terrible relationship with the cinematographer, he tried to make my life hell. He had a personal vendetta and was so juvenile in his approach. I found out later he had tried to get his friend who was a sound mixer on the film, but I had a 5-time relationship with the director so... I was an out-of-town hire,

I've always prided myself in that, but, you know, you can never be 100 percent.

Edgar
I would be interested to know in terms of directors, there are some directors that are more interested in sound and there are those who are not so interested.

Does that change the approach when you work with directors that, for example, I don't want to say that they don't care about sound, but it's far down on their list of priorities.

Sonex sound baffling put inside a concave ceilinged church for live singing on *West Side Story* directed by Steven Spielberg - 2019

Tod

Yeah, of course.

It changes it more in my interaction with the director.

If I realize that they really don't care, they just want me to get the sound, then that's what I'm going to do. I am, as I always am, going to do my job the best that I can, and I have less interaction than I would like to, that's just the way it is.

I'm not going to change them, so I'll just do the best job that I can possibly do.

Hopefully, when they get into post-production and their sound mixers in post-production are very happy with my work, maybe that passes on.

I like working with good people, to the point that when I go in for interviews on jobs, I interview the director as much as they're interviewing me. I want to know what their feelings are about sound and I also want to know what kind of person they are and what kind of set that they run, is it a pleasant set or is it going to be a miserable set.

There's only one director I worked with that I really disliked working with, he didn't care about sound and he ran a miserable set. I try to feel that out beforehand, so I don't get myself in that situation again.

40

In your movies, do you always manage to work with the audio post-production team? Can you follow the entire process of sound post-production on your movies?

On most films, they haven't even hired a post-production sound team while we're filming. If you think about it, they edit the film for four months before they bring in the post-production team.

A lot of times we won't even know who the post-production team is going to be.

I am very much on my own to figure out what the best style and the approach to record the film. When we do musicals, everything changes because now there is a post-production team hired and ready.

For instance, *West Side Story* was great because they had two of the best post-production people in the business, Andy Nelson and Gary Rydstrom and we were in contact four months before filming began. We even wrote Steven Spielberg a letter as to why we should do live vocals with a boom microphone, and if he needed to paint it out with visual effects, that it would be worth it to do that.

Steven had never done a musical before, so we had this great opportunity from the three of us, it held some water. By the way, Steven didn't go along with painting the boom out but did with a lav if we really needed it.

I always wished that post-production and production communicated prior to filming, but the reality is, it's just timing, if Post sound isn't going to be starting for another four months, there's no way that they can hire them while we're filming.

So, I understand that side of it. I always just try to give them as much material as I can, and I take notes on everything I give them.

We put out extra microphones for everything, on my metadata I'll type in this channel is "kids playing hopscotch" over here, this is "a car goes by over here".

I cofounded the Hollywood Edge Sound Effects Library, so I know how to label.

If it's a period film, I'll always take the main cars that we use in the film and do a little sound library for each one of them.

For the movie *The Irishman* we had like three primary cars.

We took like two days and did a whole library of them from the inside, outside, doors closing, starting the engine, driving, stopping, everything.

This way, the more that we can give Post, the better the movie sound is going to be. That's my ultimate goal.

I've gotten work, through post-production people because they're happy with my work, and I think that vice versa is also true.

People that don't turn out great work that I've heard about from many post-production mixers, that this one he doesn't do this, he doesn't do that or whatever, and that's a real detriment.

It's not going to get you hired again.

So, more is more and more is better.

Guntis Sics
Australia

44

Guntis Sics

Guntis Sics has been recording sound for over 25 years, covering a wide array of projects. His career began at ABC in Australia where he worked on everything from children's shows, news and current affairs and finally to drama. The freelance world beckoned and then he commenced working on feature films and TV drama. What he loves about the sound field is the combination of creative energy, operational complexity and cutting-edge technology.

And of course, the people!

He possesses both an EU passport and an Australian passport and has recorded sound in over 45 countries to date. He is based in Sydney.

46

On location for *Ticket to Paradise* directed by Ol Parker - 2021

What do you like the most about this job and what makes you the happiest? What are the pros and cons of the life of a production sound mixer?

For me there are three dimensions to this job: the technology, the logistics and the psychology involved in dealing with the wide variety of people and crafts.
At any given time, at least one of those things gives me joy.
Sometimes it's the technology: the miracle of digital recording, the quality of wireless systems, the incredible portable recording systems we all conjure up; at other times I really enjoy the logistical issues, the complex dance of interacting with cameras and actors and where to place a microphone, how to cover the dialogue which moves through space and changes with every take.
Finally, the psychology of a film set, or perhaps the sociology, but essentially working alongside people with a wide variety of skills, from actors to grips, from designers to editors.
It is fascinating to observe the myriad talents on a film set and a great privilege to work alongside them.
As for the pros and cons of life as a sound mixer, well, that's a very big question which is impossible to answer in all but the broadest terms. Basically, the pros outweigh the cons, clearly… otherwise I would have stopped a long time ago.
There's a great satisfaction with working on wonderful stories and being a key contributor.
I personally also love project-based work, rather than working on the same job year after year, the variety keeps me interested.
Of course, there is the travel and associated benefits, seeing the world, meeting people and exploring different cultures, truly one of the best things about the job.
On the con side, the biggest thing would be missing my family and friends, followed closely by the long hours we work. I've never liked that.
Sometimes the inconsistency of work is a negative but as my career developed that seemed less of a concern.

47

If you had to explain your job on set to a child person? who loves the Cinema, what would you say? Would you encourage them to pursue this career?

I would say that whenever an actor talks, I'm there to do my job which is to record that dialogue.
It's a simple as that.
If you are at the movies and you can hear people talking, that's down to me.
I would encourage them for sure, sound/audio is a noble, wonderful and intriguing craft/art and has a colourful and amazing history of invention and development.
I'm proud to be part of that, and I think anyone would be.

You have worked on many internationally acclaimed movies. Were you ever afraid not to be up to the task? How does one reach such levels in this job?

The first thing I have to say is that I've been lucky: in a way I fell backwards into the job and have been blessed by circumstance many times.
Having said that, I have always been rather single minded in pursuit of my dreams, by which I mean I never gave up, I never caved in and I never seriously considered doing anything else even when times were tough. And they were: I went broke, I lost friends, relationships, missed birthdays, weddings, funerals, didn't get jobs I really wanted, basically all the pitfalls one could have.
I think that determination and on the flip side, the ability to roll with whatever is happening is the key to success.
I have to say also that opening yourself to whatever the universe throws at you is a useful character trait, along with not feeling sorry for yourself if things don't work out.
Also, I was very lucky with my first years in sound at the ABC (in Australia), which was a fantastic learning institution: things were done properly and thoroughly, our boss was extremely knowledgeable and ran a very unusual department in that we all felt like we were friends rather than workers.
It was very common to spend extra time after our shifts just hanging around talking about sound simply because it was enjoyable.
This type of grounding meant that when I left the ABC, I was very thoroughly trained so my confidence was quite high.
I've never really felt as though there was a situation that I

48

couldn't handle, obviously there were times when I struggled but, with strong fundamentals, I always knew I could fall back on, there was no problem too big.

The only thing that throws me sometimes is the psychology of a project, when there are people with huge egos or misguided ideas about their self-importance.

That can be hard to cope with, but generally a smile and a shrug will do it.

When you record sound for a movie, what is your main objective? What do you try to provide to the post-production team to ensure a scene is complete (from the point of view of the sound recorded on set)?

The key ingredient for a soundtrack (most movies anyway) is dialogue.

I endeavour to supply as many options for the dialogue as I can, so on the boom mic, on a lavalier and whatever else might be required, all on discreet tracks so Post can find the best version and use that.

I also try and keep the background consistent, so the editing is easier.

I have to say that my singular focus is on the dialogue, but if I have time and opportunity, I will record anything else I can: atmospheres, fx, wild tracks.

I also try and record in surround and stereo anything I think suits that, I would never compromise on the dialogue in favour of anything else though.

What makes you disappointed/dissatisfied with the sound you are recording?

I've long since stopped worrying about things I can't control, so any disappointment I feel is only ever about my own performance: if I miss a cue or I haven't chosen the right microphone, whatever it is, it's down to me.

As much as we endeavour to control other elements, I never take it personally if a plane flies overhead, I don't agonise if an extra is making noise, I just deal with it as it happens.

In my early years I did take it personally, there were days that I felt tortured because the sound I was getting wasn't up to my standard, events were conspiring against me and it seemed like there was nothing I could do.

On location for *Extraction* directed by Sam Hargrave - 2019

50

Over the years, I've become much more philosophical about it, to the point where I can now clearly recognise the things I can control and the things I can't.

This gives me the freedom to blame myself only and not complain about the things that are not under my auspices.

Having said that, I do always feel disappointed if the results are less than excellent, I mean it's in my nature to want perfection, so there are days when I feel I could have done better. Mostly those days are due to my lack of planning, which is down to me.

Can you give us an overview of your professional tools/ equipment? Tell us what they looked like in the past, how they are today, and if you have one piece of equipment has always followed you on set in all your movies.

The tools of the trade have changed enormously over the years: we've gone from mono recording on 1/4" tape to digital multi track recording.

My career started with a Nagra 4.2, which was a fantastic machine, engineered perfectly to work in all conditions and very reliable.

In those days, we were very disciplined about how much we would shoot because we all had to stop to reload every 20 minutes or so and this interruption helped to focus everyone on what was necessary and not just keep shooting non-stop. I carried that Nagra around to over 40 countries and it always worked even in the worst conditions.

Nowadays, on feature film, I have far more equipment than ever before.

Sound mixers are basically mobile multitrack recording studios combined with a broadcast station. The way sound is delivered these days means I need a much larger mixing console, a multitrack recorder and then a ton of wireless equipment.

We used to just record what one single camera was seeing, but now, with multiple cameras and the editing requirement, we record the whole scene and all the actors no matter if they are on camera or not.

The best change has undoubtedly been the huge improvement in wireless lavalier systems: early in my career you had to almost be sitting at an actor's feet to get a signal, but these days it's normal to have clear audio from very long distances. Microphones have improved as well, though not as dramatically as wireless systems.

As for mixing consoles and recorders, they are a far cry from the Nagra and these days my whole system is digital and includes Dante networks; there is barely an audio cable to be seen.

I have a Cooper mixer which has been in my kit for 25 years now and even though it is no longer my main mixing console I can't bear to part with it and still use it from time to time because it just sounds so good.

It's worth noting that, in my experience, when technology goes through a massive change, audio quality dips at first until the new technology improves.

Basically, we follow technology mostly for perceived convenience rather than quality improvements.

The first DAT recorder I owned sounded awful, but you didn't have to change reels every 20 minutes, so we used them anyway.

Now, at last, the digital recorders actually sound fantastic.

You have worked with many famous actors. Often, in order to capture good sound (especially with wireless microphones) we need to stay really close to the actors,

place the lavalier, change positions and so on, or other times we need their help by speaking louder or doing other adjustments. How does one manage these dynamics with the actors?

This is a very tricky part of our jobs, in some cases the hardest thing of all.

To begin with, the vast majority of actors are really lovely people and are happy to help in any way they can, you can make adjustments as often as is needed and they don't complain.

I think most of them understand that it is in their interest to get good sound.

Having said that, there is a type of system for bigger movies that puts actors into a bubble and makes access to them a lot harder.

Many famous actors only appear on a set when cameras are about to roll, so we use stand-ins to set the shot up and rehearse it and then at the last minute the actor is put into place.

In the end, this makes it hard to establish any kind of working relationship with an actor, when you never see them except in performance mode, and then when cut is called, they are rushed back to their trailers.

The end result of this can be a complete absence of any connection, where often the costume people even put the lavaliers on in the trailer and we don't get to hear anything until they turn up on set moments before cameras roll.

Of course, this is a very awkward way to work for sound and it can be difficult to get quality.

I always give the costume person some headsets so they can actually hear how the microphone sounds, and, if they have to adjust it, this really helps them to understand the various issues.

Some of this is understandable, acting can be very mentally challenging and being interrupted for technical reasons must be frustrating.

We try extremely hard not to mess with actors unless it's really necessary but, of course, it does happen.

One thing that has changed in recent years is the ability and desire of picture editorial to do quite a sophisticated mix of what we are shooting very early in the process.

Well before sound post arrives, they are mixing tracks together, cleaning up background noise and
showing high quality dailies.

This all has a flow down effect back to us on set where the dialogue recording becomes more important and most directors will now get at least one good clean version of the dialogue which can be used for dailies.

Sometimes if it's a big fx shot with wind machines etc., the director will ask to shut it all down and shoot one take without the noise. Actors too are noticing this and adapting, and their desire to get good sound is usually quite high.

Wireless microphones have become an essential tool of this job. Their proper functioning is linked to the clothes and costumes that actors wear. How do you interact with the wardrobe department to make sure there are no problems? Also, do you have any real-life examples you would like to share regarding superhero costumes?

The wardrobe department is a critical relationship as far as I am concerned.

Putting lavalier mics on actors is always awkward and we need help from the costume people every single day.

Some of them know how to put microphones on better than sound people, which makes sense given they know the costumes so well.

But it's also about access, and the more prominent actors have their own people and prefer it if they do the microphone placement.

This can work well but it also can be difficult because if it sounds bad sometimes there's not a lot we can do about it.

In the end, putting a microphone on someone is a fairly invasive procedure and highly intimate at times, so any help we can have we are grateful for.

On *Thor: Ragnarok* the leather vest Chris wore was very squeaky at first.

Over the weeks of shooting, it softened up and began to sound a lot better, but the problem was it also started to look worn so of course it was replaced with a new one… back to the squeaking I'm afraid.

How many people make up your standard sound team? Which characteristics should a good boom operator have? Do you personally place the wireless microphones on the actors?

54

On location for *Operation Buffalo* (TV series)
directed by Peter Duncan - 2019

These days a standard sound team on a reasonable size feature film is four people.

If it is a musical then more crew are required but given that we record even the off-screen lines and supply mix tracks and ISO tracks and comms as well, 4 people are required.

Two of them are also boom operators as I often use two booms all day long.

A good boom operator needs a very particular set of skills: firstly, the kind of spatial sense which helps them navigate the lights, the camera, the actors... all of which can be moving; secondly, a good ear for dialogue and it's rhythms and dynamics; thirdly, a good memory for words and actions, and fourthly the type of saintly patience which all sound people need.

Many years ago, I recorded a film in Beijing, China, which was all in Mandarin.

I don't speak Mandarin and neither did my boom operator (Fiona McBain).

Much to my surprise, she managed to somehow feel the dialogue and unerringly swung the boom to whoever was talking. I think language is universal in that sense, it has a rhythm and a dynamic that no matter what is universal; she was able to hear the intent of it without understanding a single word.

Regarding the second question: I never put the wires on actors these days, someone in my team or someone from costume does that, I just listen and comment.

Have there been times in your career when you could not secure a usable sound – and how did you react when there are unsurmountable difficulties? Any interesting anecdotes you would like to share?

Of course, there have been many times I could not secure usable sound, I've been recording for 40 years after all.

It's probably helpful to define what you mean by usable, because that's a critical thing when it comes to recording sound for picture.

For example, I've recorded a scene where the part we shot in the morning was usable, but the part we did in afternoon wasn't.

Same microphones, same setups but the background sound had completely changed so it was impossible to cut it together without ruining the sound overall.

So, usable really means can the editor and sound post use it, rather than does it sound OK to me. By contrast, I've had occasions when I've been very unhappy with the sound, but it has been used anyway and worked really well, so again, the point is that usable is a very specific definition and often unrelated to the actual quality of the recording or what I might personally think about it…

Having said that, there are times when nothing you do will save a recording, mostly because of the physical circumstances, for example rain or wind or background noise we can't control.

On *The Wolverine*, I remember we couldn't get usable sound at one location because there were helicopters all around us doing training runs and they wouldn't stop.

Usable also means appropriate, in the sense that if we cannot see something then we don't really want to hear it.

So, if we'd seen those helicopters, it would have been OK.

For example, a highway noise nearby while shooting a period piece…

I had this experience in Taiwan during *Silence*: the 6-lane highway was right next to the backlot and I could hear it very clearly. Luckily sound post were able to clean it up enough so the sound was usable. I did a trial recording before we went there and sent it to sound post to test. It was usable.

In your opinion, how much should a production sound mixer "fight" for certain things, and when should one accept the inevitable compromises? Tell us more about your relationship with other roles and departments such as directors, camera, wardrobe, cast and please share any anecdotes you might recall.

First of all, I'm not sure I like the word "fight", but I suppose that sometimes it does boil down to that.

I would hope that diplomacy and clear explanations work well enough but sometimes they just don't.

The perception of production sound varies quite a lot, not only between departments, but also countries; I've found that here in Australia there was, and continues to be, a culture of dismissal when it comes to recording on location.

There is a long history here of film schools not really bothering too much with production sound and concentrating on post-production. This has had the inevitable effect of generations of filmmakers thinking that it's largely impossible to get usable sound on location, and then allocating miserly resources to it.

Often, they are really surprised with the results a profession-al hard-working sound mixer can deliver, but the attitude is sadly quite ingrained.

Producers vary wildly between those who insist on good sound and those who think location recording is a waste of time and money, with the latter outweighing the former.

If I have to fight for anything, I often find that my closest allies are the picture editors, who can be very grateful to get good sound, which makes their job a lot easier.

It's a rare moment when I will say anything much on a film set, because by then it's often too late and nothing much can be done.

The time to fight is mostly in pre-production, about the sets, the locations, the generators etc. Once these are all in place there isn't much left to fight about usually, and honestly if a director doesn't want to do another take, I'll give them the benefit of the doubt that they are 100% aware of the conse-quences.

What does get under my skin is the conflict of interest my department has with other departments, in particular the ADs, who can be quite careless about the sound.

I like to point out to them that actors care about the sound as do editors, but I suspect that because an AD will be long gone before sound editing commences, they have little mo-tivation to care whether I get a clean recording or not.

Once again, it's down to preparation, making sure everyone's expectations are aligned, and then standing firm on the day.

How will the jobs of production sound mixer and boom operators change in the future?
What trajectory are we on? From a technology stand-point, among other things.

That's a very interesting question.

I can't see filming happening at the feature level without a sound team, but I can see how we could operate remotely in the future, at least partly.

Already I've been thinking of wiring studios with Dante or similar so that, in theory, the production mixer could stay in the one place whilst mixing.

I mean this is all very easy now and the system I use is slight-ly geared towards a separation between on set and the re-cording zone.

I recently converted a caravan to a mobile recording studio

for a movie due to Covid concerns, and there were days when I simply did not go onto set at all.

I know there are methods existing and being developed which take this idea even further, and the recordist may be in a different time zone, also there may come a time when on studio pictures, the sound is actually recorded in editorial directly.

The Internet has made all of this possible, and music technology has been using it for years.

I can see film following that pathway as well.

Of course, placement of microphones will always require an on-set presence so the boom operators will have jobs, don't worry.

Another thing that may happen and has to some extent already is the use of transmitter packs on actors that also record.

I've got a couple and it's a very handy thing to have in that you can send actors off to other units or for pickups somewhere and record them without being there.

Perhaps one day every actor will have a wire and a pack all set to record, all coded up and with smart limiters and auto-mixing, and someone's job will be to compile everything at the end of the day.

Recording without listening.

In the meantime, the trajectory we are on is technologically very exciting as far as I can see, better wireless gear, better multitrack recorders, better everything, so we can do a better job.

On location for *Ticket to Paradise* directed by Ol Parker - 2021

Chris Munro
UK

Chris Munro

Chris Munro is a production sound mixer who has collaborated with some of the most respected film-makers including Steven Spielberg, Ridley Scott, Paul Greengrass, Ron Howard, Alfonso Cuarón and M. Night Shyamalan on worldwide locations for both UK and US based productions. He has won two Oscars and two BAFTA awards, winning 2014 Oscar and BAFTA awards for the ground breaking space adventure *Gravity* and nominated the same year for *Captain Phillips*. Previous nominations for *Backbeat, The Mummy, United 93* and *Quantum of Solace* winning BAFTA award for *Casino Royale* and an Oscar for *Black Hawk Down*.

On Steven Spielberg's *Ready Player One* - 2016

When did you realise that your professional life would involve sound? What were the beginnings of your career?

This is quite a difficult question because I started when I was 16 years old. I had two great interests: one of them was the cinema and the second was electronics.
I used to go at least a couple of times a week, every week, to the cinema at first with my mother and then on my own. We also visited the library every week.
My mother was a very big reader and took out as many books as possible. I would look around the library waiting for her and started to find books and magazines on electronics and wireless construction, which was a big hobby at that time and is how I started to get involved in electronics.

This was in the 1960s, when electronics were changing from valve technology to transistors. Getting involved in transistor electronics, and making my own electronics projects led to my interest in sound. However, my main interest was cinema, where I always wanted to work, or perhaps in television. In those days it was impossible to get into that work because it was unionised. You had to be a member of the union, and if you were not a member of the union, you could not get a job, and if you did not have a job, you could not become a union member. Are you familiar with the term a "catch-22" situation?

My third love was cars, so when I was about 15 years old, I used to work in a petrol station and mess around with cars. One Saturday, a gentleman came into the garage to fill up the car with gas but then he couldn't start his car. It was a common fault, which I helped him with. He very foolishly told me he was on his way to the studio, and I realised that he worked in films. As was common in those days, he gave me his business card and I then drove him crazy to find me a job. He told me there were no jobs in the studio, the only job was in sound and you had to be somebody with experience in electronics. I said: "Well, I do". Of course, he didn't believe me because he thought I only knew about cars but arranged for me to apply for the job. I went to the studio to be interviewed by the head of the sound department. He was an older gentleman who knew nothing about modern electronics, but more about older valve electronics. I knew more about the questions that he was asking me than he did.

65

Anyway, to cut a long story not quite so long, he offered me the job. The problem was I was only sixteen years old, and I should have been going on to my further education. I told my parents it was a summer job. I started there, I worked all summer and I kept going to work when my parents thought I was going to school until a few weeks later when the school called up to ask why I hadn't come back. Although I started in sound it was not my intention to stay. I just wanted to make films. Sound was a way for me to start, but I guess I must have been quite good at it and progressed quite well.

I have always made my own films that I have written and directed but I'm best known for sound. In fact, I recently started an ethical production company with my business partner who is the UKO of Google in UK and Ireland.

Edgar
When you started working in sound, did you start as a boom operator?

Chris
This is another funny story. A boom operator is a very skilled job, you cannot start as a boom operator. I started as a maintenance engineer looking after the editing equipment. In those days, I used to wander around on to the different stages talking to the different sound guys. When they needed additional help or someone to hold a microphone, for example, they would let me help.
One day, I was asked to go on to a set where the boom operator had had an argument with the director, left the set and them without a boom operator, so I took over for the day. Next day when I came back in, they had hired another boom operator, but the director said he didn't want this other boom operator, he wanted me to continue because I didn't give him any trouble. That was how I started boom operating, as well as doing maintenance, and after that I went freelance, I had worked on a lot of films with Roger Moore, at Elstree Studios where I started, but he was going to Pinewood studios, to work on a series with Tony Curtis and Roger Moore called *The Persuaders*. I was asked to go with his team and worked on that for some time. Later, I worked as a freelancer boom operator but also doing some maintenance which meant that I got to go on a lot of locations where maintenance was very important, where I had to look after the equipment.

About 1975-1976, I decided that I wanted to stop being a boom operator and become a mixer. I was offered a job to work as a second unit mixer on a film directed by Sam Peckinpah, *Cross of Iron*, and that was my first film as a sound mixer.

Looking over your filmography, it is impressive how many world-renowned films you have worked on. How does one reach such levels, what characteristics should a production sound mixer have to reach and maintain such levels?

You have to remember that I have been working in sound for over 50 years, so you would expect to have quite a few films in that time. First and foremost, you must love films, and so I always describe myself as a filmmaker who specialises in sound, rather than a sound person who specialises in films. I've always wanted to work in cinema, and I have tried to pick films that I thought would be interesting and would be good films to work on.

Edgar
How did you reach these technical skills and how did you keep them over time?

Chris
Because I am from a technical background, I have tended to do lots of very technical films.
I did five James Bond films and three Mission Impossibles. I tend to do films which have a lot of technical input, *Gravity* for instance, was an extremely technical film, there was a lot to work out. I have always been interested in technical challenges and they are the kind of films that I get offered. I think it is something that you get known for, and obviously when you have had success then you get asked to do other films. I've always been someone who likes to meet new challenges and was one of the first people to start recording digitally, I was very interested in digital technology, and I still am. I'm always looking for new things, new pieces of equipment and new ways of working. I was probably one of the first people to start using multi-track for recording. *Black Hawk Down* was shot digitally and in multi-track when most films were still being shot in mono. I was shooting in eight track and sometimes in 16 tracks on a digital recorder.

I worked for many years in developing equipment, working mainly with Fostex, I now work quite a lot with Sound Devices. I am very interested in new technology and in developing technology. I always try to seek improvement in all of the films that I do and never sit back.

I try never to do exactly the same things I did on the previous film. Precedent is often a thing of filmmaking: they look to see how other films are made and then try to do something in exactly the same way. Whereas I always try to advance to the next stage.

Edgar
It is really incredible to hear you say this, because one might think that after all you have achieved and all the awards you have received, you could just sit back and relax and do your thing. Instead, it is really refreshing to see how much curiosity you have and how much you're eager to always do something new, something more.

Chris
Sure, that's a problem. I don't like to sit down.

When you look at this profession, what are the most challenging aspects of this job?

Every film is a challenge. For example, there are cases when the actor doesn't want to speak up or whose performance is such that it's very, very quiet and it's difficult to record. And we need to find ways around that challenge because our job is to capture the performance and not to influence it.
So, we have to find ways to make that very low voice recordable and to be able to give a performance without influencing it, so that's one challenge that we often have.
Another challenge may be a location, which can be very noisy, so we have to try to find ways of making that location work for sound.
Now, bearing in mind that, you know, lots of locations that directors pick, they pick them for a certain look or for a certain reason. Our job is not to go, find a location and be negative about it, saying: "Oh, this location is terrible, we can't shoot sound here". Our job is to try to make that location work for sound. Part of it may actually be discussing with the directors to see whether it really is as great as he or she

thinks it is, and perhaps even convincing the director to go to other locations. Or perhaps convincing the director that maybe we could shoot part of the scene where we are looking at, whatever the noisy thing is, perhaps a railway track or whatever, and then shoot the dialogue somewhere else.

There are always difficulties; unlike on camera, where you can see all of your problems within your frame, the problem with sound, of course, is that most of our problems are outside of our frame that we have to deal with, and that is always a challenge.

There are countless other obstacles, but I don't think of them so much as challenges, I think of them as part of the job, and it's part of what makes you do the job and it's part of what makes you good at the job. Experience is crucial, because mostly you have encountered these issues before and you normally have solutions. This way you can often look at things calmly and try to evaluate what is important and what is not.

One of the most important things, in my opinion, is to give the filmmakers confidence in what you're doing, help them know what is usable and what is not, and understand how a film is cut so that they can know what parts of sound are important, which ones are to worry about and what can be fixed in posts without any detriment to the film, for example.

So yes, our job is always a challenge, our job is full of challenges but that's what we do.

Based on your experience, do you believe there is an "European" or "American" style/approach to this profession? Yes? No? Why?

Yes, there are certainly different ways of working, and I guess in the UK we work more like the Americans than the Europeans because we share a common language and most of the films that we make in the UK are in some ways connected to American studios, so I guess in the UK we're much more like the American model of films.

European films are different in many ways, and it is why we love European films.

One of the things that is very different is that it is very common for a sound mixer in many European countries to do the whole process from the beginning of pre-production to production and through to post-production. So, for example, they will be the production sound mixer, they may be the editor, they may also be the re-recording mixer, it's quite common in Europe to do that.

I have owned a post-production company for about 20 years, so for one period in my work I sometimes did that, although it wasn't really very practical in the way that we work, because it meant that you're shooting one film and then you go off to do post-production and what happens is you lose your crew… it is very hard to keep a workflow going in that situation.

And over time it has become easier in the UK, like in America, to specialise in either production or post-production, and I have tended to specialise more in production, although I do have post-production experience.

Edgar

So, do you believe that in order to have the best results, it is preferable that the production sound mixer works in production and then someone else goes on to do post-production? Is that in your experience what you have seen?

Chris

Yes, it is. I think there are many advantages in being involved in the whole film, but, you know, sometimes it's good to have other ideas to be able to see it from different directions, and sometimes that's what worked very well in specialising.

It also means that it is easier to keep working with the same team, because what tends to happen is that if you're doing production with one team, then you lose that team to go do post-production with another team, then when you go back to production, you lose that post-production team.

We're not alone, a sound team is a team of collaborators.

For instance, a sound mixer is useless without a great boom operator and a re-recording mixer is useless without a great sound editor. So, in my opinion, I actually think the best results probably do come by specialisation, although I do understand the benefits of seeing a film from start to finish because you don't have to do detective work for instance, you already know where everything is, how things work.

And so that's why it is very important as a production mixer to keep very good metadata, to keep very good control of your tracks so when you hand them over to somebody, they know exactly where everything is, how to find stuff, how it fits together. We are in the habit of doing that, and we tend to often work with the same post-production teams, so they know how we work, and we have a kind of common language between us.

On Christopher McQuarrie's *Mission: Impossible - Fallout* - 2017

After working together so much in this way, it is almost as if you were doing the whole thing yourself but using the best of everything available.

When you record sound for a movie, what is your main objective?
What do you try to provide to the post-production team to ensure a scene is complete (from the point of view of the sound recorded on set)?

We have already mentioned how important the metadata is, and how keeping it well organised is vital. The first thing that we always concentrate on is the dialogue: the most important thing is that the dialogue is clean, clear, well-recorded: meaning that we can actually understand what is being said. So, for example, I never work with a script on set, I never follow a script on set, I always listen to make sure I am understanding what the actors are saying. I understand the performance and then I capture the performance.

The performance is very important to me, we have to understand the performance and that is where we blend our skills of technicality and creativity. Although there is a technical side to our job, the main part of it is creative: it goes without saying that we need to have those technical skills, but what makes a great sound mixer on set is somebody who can bring creativity to that. And so, the point is that we often will make sure, for example, that we don't already cover the dialogue on camera, we cover the off-camera lines at the same time, we try to capture all of the breath and emotions because these are all the things that make a performance real, not just the words. Dialogue and performance are always our main objective but we will also cover other things at the same time, for example, we will always try to have and to record background noise and background atmosphere. I like to try to do that now that we have multi-track capability, I like to try to do that while we are recording the dialogue. So, I will put microphones away from the dialogue, away from the scene, recording just the background noise, meaning that it makes it much easier for the editor and sound editor to cut between takes. For example, if there was a motorcycle that passed on a really great performance, we've got that motorcycle clean so we can carry that on to the next cut on the close-up or whatever of the performance. Similarly, we are able to use a take where perhaps there isn't a motorcycle.

We always try to record sound effects of things which are unusual, so we try to keep a catalogue of all the sound effects of things which are important to the film and are unusual for the editor. We keep a catalogue of all sounds; cataloguing is very important.

When recording, what makes you happy and satisfied and, on the contrary, what causes you to be disappointed or dissatisfied?
How important is it to plan, organise one's work and foresee as many difficulties as possible beforehand?

Yes, I think planning is vital. Planning is very, very important and we plan very much.
Nowadays we tend to put a radio mic on every actor and we do that not because we like the radio mic sound better than anything else (we always prefer the boom sound) but we put them because sometimes there is something that can just be captured, whether that be a breath, a word we weren't expecting or whatever. One of the parts of our preparation is working closely with the costume department and with the actors to make sure that we can radio mic them efficiently, and that our radio mics work well.
Other planning includes just looking at sets and making sure that, for example, if there's a hard floor or there are very poor acoustics, we try to make those acoustics good before we shoot there, so we might suggest putting a carpet down if it's a very hard floor before we start shooting, and again, preparation, we work very closely with the other departments.

Often problems we have these days with lighting is a lot of lights have fans in them and make noise, so we work very hard to make sure that dimmers and generators and all of those sorts of things are away from the set.
Asking me what disappoints me most in the sound it's probably when something destroys the sound or makes the sound not as good as I'd like it, when it might have been unnecessary, for example, when someone just insists on putting a generator too close and that has an effect on the sound or for a piece of equipment that is not a part of the film that happens to be put on set without people thinking… these are the things that make me upset or angry, things that we could have controlled; the others aren't.
I think we all have a responsibility for all aspects of the film,

and to get good sound on film isn't just the responsibility of the sound department, it's the responsibility of everybody in the same way, it's the responsibility of everybody to produce a great film. The most important thing is that we all collaborate and work for the same end product.

We're accustomed to what sounds good and what sounds good is what sounds natural, what captures the performance, that's what sounds good and of course, like anything, if we don't get what we think sounds good, we're disappointed.

In your opinion, how much of an "artistic" side and how much of a "technical" side are there in the work of a sound engineer on set?

I think our job is wholly creative, the technical side is something which is a given, everybody has to have that skill.

So, for example, you are working as an interpreter, you couldn't work as an interpreter if you couldn't speak English, that's part of the job, but what makes you a good interpreter is if you can understand what is being said, you may specialise in a certain kind of interpreting but the baseline is that you have to have the knowledge of English.

So, for us it's the same thing, to be a sound mixer the very first thing you must have is technical knowledge, that's a given, that's just a matter of speaking the language.

But what you also have to do is have to have an understanding of performance, have to have an understanding, a vision of how a film should work, how it should sound and a creative interest so when you can talk to the director, you can interpret what he or she is looking for.

Creativity is the part which separates a good sound mixer from a not so good sound mixer and not the technical side.

How many people make up your standard sound team? Which characteristics should a good boom operator have? Do you personally place the wireless microphones on the actors or is it responsibility of your boom op/sound assistant/utility?

The standard team in the UK is a little bigger a little bit bigger than it is in Europe, we have now in the last, I guess, 10 years, always had two boom operators.

We have two boom operators for a couple of reasons: one is because very often there are two or three or even more cameras, so it is important to be able to cover more of the set, so two boom operators most of the time. Filmmaking has changed and it really doesn't work if you have a boom moving between actors all of the time because you miss the small nuances, the breathing, the different parts of the dialogue, which aren't necessarily words, so we always try to have two booms at all times.

We also often perhaps have another boom covering the dialogue which is not on camera, so most shots will have at least two booms, some will have three booms, but two booms are standard.

We then would have a second assistant. Now different people work different ways: on my team my second assistant is the person that normally looks after the radio mics.

He or she will have a relationship with the actors, will work with the actors, will work with costume and will be responsible for the radio mics: they go off, they fit the radio mic, they listen to it on a portable device to check that it sounds okay before they come to set.

Now with Covid that's changed a little because often we're not allowed to actually touch the actors, so what often has to happen is that the actors have to fit their own radio mic and my second assistant will be there with them, she will explain to them how she wants them to do it and what to do, and she'll be listening to the mic while they're fitting it, just checking that when they come to set it sounds right.

I will have had some input in that, in the early stages because I would have worked with my second assistant and with the costume department to try to make sure that we have a plan for each costume as to where the radio mic goes.

Often, for example, you can hide a radio mic in plain sight without it being seen, so often we'll look at how we can hide a microphone in the costume where it won't be recognisable in scene, that will always give us the best radio mic sound rather than try to bury it deep.

I will then also have a trainee on the set. The trainee will look after things like headphones for directors and so on, will look after perhaps some timecode, they would generally look after the equipment, but the main reason for a trainee to be there is to learn. They are really there for us to pass on our knowledge to them. Generally, a trainee on a film will already have been to film school, they will already have had probably three years of training at university or film school, and they will already be trained, but they will be learning about how to work on set, how best to work on set, and they will work on a few films as a trainee before becoming a second assistant.

A second assistant, of course, will go on to be a boom operator or what we tend to call more a first assistant.

So generally, our crew is a production mixer, two first assistants (or boom operators), a second assistant and a trainee, that's the standard crew. Obviously, if you were to do a musical or other films, then you would bring other people along, perhaps a Pro Tools operator and other people, but that's the standard for a regular high-budget film.

Edgar

You mentioned that it is your second assistant that places the microphones on the actors. Are there times when you intervene and do that, or decide to make some changes, or do you trust them to do it every time?

Chris

No, I never do that really, because for one thing it would be bad for the assistant if I was going in and undermining them by changing what they have done. It's very important to give those assistants confidence for them to get it right. It's their job and it's their domain, so I wouldn't go and undermine them. I might ask them to go and fix it again if it's not working properly.

On Christopher McQuarrie's *Mission Impossible 7* - 2020

Edgar

Which characteristics should a good boom operator have?

Chris

A good boom operator, of course, will always have the microphone in the right place, a good boom operator will understand camera angles, they will understand what the frame is.

Most boom operators only have to know what the lens size is to know exactly what the frame is.

They will try to be, as you know, in the right place in the frame, they will also know quite a lot about lighting because they wouldn't want to have shadows on the set and know how to avoid that, they will also be very good at basically running the set.

What tends to happen is that they are a little bit like a first assistant director: they are the person on set who is running everything, whereas the production mixer is back at the mixer, working and speaking to them, but they are his officers on the front line, if you like, who are running everything.

So, the important thing is to have somebody who is likeable, who gets on well with people, who can get things done and who understands what is required, and you get that from experience.

We have some incredible first and second assistants in the UK, and that's probably because there is so much work here, so they get a lot of work, and they have a lot of experience.

Edgar

So, when you spoke about the beginnings of your career, you said that you also worked as a boom operator in the beginning for a while and then you moved on.

Do you think this is a normal or preferred step to go though, being a boom operator first and then becoming a sound mixer? Is that a requirement or something you would recommend?

Chris

I think the reason why it's important to have been a boom operator is because if you haven't been one you haven't had the opportunity to work with other sound mixers.

Most of the things that I have learnt, I've learnt from working with other sound mixers.

You learn how best to deal with situations, you learn lots and things become second nature.

There are things that you can't learn from books or from schools, it's just a matter of being there, that's probably the most vital thing. You also have the possibility to learn without being as responsible, and that's very important.

Let me say that when I was a boom operator, they were slightly different times because I was working with a Fisher boom – and it was a different kind of skill operating one of those, nowadays nobody uses the big studio booms anymore. Anyway, I do believe the time that I spent working with some of the other sound mixers was where I learnt the most.

Tell us about your equipment and how you have organised your sound cart.
Is there anything in your sound kit that you could never do without?
Nowadays, in 2021, does it still make sense work with cable? Yes, no, why? Can you give me some examples?

Some very good questions. The first thing I need to mention is that I have a sound company called "Sound for Pictures Ltd", so I have a lot of equipment.

I supply equipment not just to my film, but I supply equipment to other sound mixers as well, so I do have a lot of equipment to fall back on and I don't always use exactly the same equipment. At the moment, my preferred piece of equipment is a Sound Devices Scorpio, which has 32 tracks, and I use that with a Sound Devices CL 16 fader panel. I like that a lot because of the way that it works.

One thing that is very interesting about my sound cart is that I do not have my radio receivers always on the cart. I usually put my radio receivers close to set, and I use Dante to connect them to my cart, so I try to keep my cart as small as possible so that it is very manageable. This means I can be close to the director, I can be as close to set as is necessary and not be so big that it's hard to get on set or take up too much room on set. So being small is very important for my cart, but I do have a separate cart which has the radio receivers, which goes very close to set.

Dante is something I'm very keen on, it works very well, it means that I only have to use one Cat5 ethernet cable to connect them.

Another piece of equipment I like quite a lot is the Cedar noise suppression: I do not use it in a destructive way, I only use it on tracks that I also back up unprocessed.

I find Cedar very useful for cutting out some fan noises and other noises on set, and I find it is a good way of knowing what is achievable in post-production rather than making final decisions.

The Cedar is effectively a plug-in, it's 8 instances of plug-in that I use, so that way I can use Cedar on eight input channels, and I normally do that on channels 1 to 16, while on channels 17 to 32 I select the same inputs but without Cedar. That way I'm always recording unprocessed ISO tracks,

I record ISO tracks that are both processed and unprocessed. I am currently working with Sound Devices to lock those channels together so that we can group channels 1 to 16 and 17 to 32 together. This way we can fade, we can pan channels 1-16 to the left mix and channels 17-32 to the right mix: having the channels grouped means that we can record with those ISOs a processed mix to the left and an unprocessed mix to the right.

I have previously used Cedar hardware, I have three Cedar hardware units: I have a DNS 1500, which is the one I probably like the best because it is completely manual and it means that I am very adept at using it from my post-production days and I can use it very quickly, I like it a lot, but it is only two tracks. I also have a DSN 2, which is the small Cedar hardware which I'm not so crazy about, again, it's only two tracks and it's automatic, it's great but I can do a better job with the DNS 1500.

I also have other hardware, one new version which is called a DNS 8D that uses Dante and gives me eight instances of Cedar, and now works very well because that gives me a lot more control, giving me the best of all worlds because I can have some manual control for each channel, plus, because of Dante, it's very easy to connect.

I am split between them because sometimes I use that and sometimes, I use the software version which I've recently acquired for the Sound Devices Scorpio.

The good thing about the software version is that, of course, it doesn't make my cart any bigger, it's small, but it doesn't have the same amounts of control that the DNS 8D has.

One of the advantages of using eight instances of noise suppression is that we can actually use different degrees of suppression on each channel and that's very important because very often the booms, for example, will need more suppression than the radio mics or a boom which is closest to the noise source.

It's also better, I think, to use noise suppression on the ISO tracks rather than on the mix, because if the editor needs to find ISO tracks, he already has them available with noise suppression, and I find I can do a better mix that way.

But remember, the most important thing is I'm not doing anything that cannot be changed.

I'm not doing anything that is destructive, when the sound editor gets my material, it is completely raw, he can completely ignore my Cedar tracks if he wants to do.

Edgar

Regarding booms and radio microphones, which equipment do you use?

Chris

I change microphones all the time, but generally I always come back to the Sennheiser MKH 50 and MKH 60, I like them for dialogue because I like the fact that at different distances, they still sound similar.

I love the Schoeps microphones when they are in exactly the right place, but I really dislike them when they go out of range, and I don't like the way that the characteristics change with distance.

I very much like the MKH 50 and 60 because they sound so consistent in where they're placed, so they're my usual first go to microphones, but I have to tell you that I carry a lot of microphones.

For radio mics I very much like the DPA mics, I like the DPA 6060s, mainly because they are so small, they sound great.

I used to be not so keen on the older DPA microphones because, a bit like the Schoeps, if they weren't in exactly the right place, they just didn't sound quite so good, but now I'm very keen on them. For radio mic systems I use Audio Limited A10s and A20s, I like those a lot.

And you ask me a question about cables: well, I always insisted on cables on my booms until quite recently, but now with the A10s, especially for booms, I am very happy to use those on our booms because I think they sound great, I think it's very hard to notice the difference between those and the cable, and it gives me the advantage of using the Dante ethernet connection for everything. This means that I have fewer cables overall.

The A10s and A20s, in my opinion, are the best sounding radio microphones system.

80

I like the A20s a lot because of the 32-bit float, which means that they have a huge dynamic range which you can use, and I also really like the fact that they record.

A very important part of my work is recording, and this is something I do, which perhaps other people don't do quite as much. For example, in *Mission impossible*, all of the car chase sequences, and action sequences are shot with cameras on the car, Tom Cruise does his own stunts, he just goes off and does them, no one is there, so it is very important for us to have recording radio mics.

What I do is: I set all the timecodes and I record everything on A10s and then at the end of the day, I re-conform them into files, so I use the Mic2wav reconfirming software, which is something that I worked with Sound Devices to develop.

Therefore, using radio mics that can record for me is vital and it is the way forward.

Today, movies are increasingly shot using multiple cameras, which makes boom operating extremely difficult. Could this profession have survived without wireless microphones? What are the most important aspects to always keep in mind to get a great sound from wireless microphones?

I think the important thing is that the radio microphones are not our main source, our booms are always our main source. Remember that we often have three booms, that is because there are multiple cameras. It is pointless if you have two cameras and you have only one boom operator. I think it's very important to try to make other filmmakers understand that.

Radio mics are not a substitute for the boom, they are in addition to the boom.

Sometimes I hear filmmakers saying: "Oh, why do you need the boom, because the actor has radio mic on; or why do you need radio mic because you have the boom". It isn't one or the other, it is a combination. The prime microphone is the boom, it will always sound best and that is what we strive for. The radio mic is something which just gives a little bit more, it is the icing on the cake, or it just gives us the ability to make the sound seem a little bit closer or maybe to improve the clarity a little, but we still want that sound of an open microphone.

Let's be clear: if microphones the size of a pen head would be as good as big microphones, why would we make big microhones?

It isn't a matter of booms or radio mics, it is a matter that both have their place, but the boom microphone will always be the most important microphone for sound mixers.

Edgar
When using radio mics, what is the best advice you can give? What is something that you should always keep in mind to get great sounds specifically from wireless microphones?

Chris
Pay a lot of attention to the level.

What you don't want to do is to have the level set too high, that you are using the compression too much on the radio mics, you don't want to hit the compressors; it is very important to try to get the best dynamic range you can get. Now that we have 32-bit float radio mics, that has overcome quite a lot, because one of the old problems with radio mics was that the dynamic range was very reduced. That is significantly better now.

The other thing to bear in mind is mic placement: take your time, make sure the microphone is placed in the right place because that is what is important.

82

Our relationship with the actors and with the director is crucial to perform our job successfully.
Both can help us get good sound and we often need to make demands of them.
How do you relate to actors and directors?
What should the right profile of a production sound mixer be in relationship with other departments? Any interesting anecdotes you would like to share?

I can't think of any anecdotes that I can share without being sued.

Obviously, our relationships with everyone are vital, especially with directors and actors.

Besides the relationship, as we said earlier, it is important that we understand performance and our job is to capture performance and not to influence it.

As far as directors are concerned, I like to work in a way that we decide very early on what the director is looking for. I try to discuss a lot about the sound with them early, so we have a common idea. They are listening to the sound all the time, but I tend to try to only get involved or to discuss the sound with the director when there is a real problem that is not solvable.

This way the director knows that if I have an issue that I talk to him about, he knows it is serious. On the whole, I try to give him the confidence that I am getting what he or she wants, I think that is very important and that is what I try to deliver.

Edgar
And with famous actors, with great actors, what is the relationship there and how do you usually approach them?

Chris
Well, there are a lot of actors that I have worked with many times, you know, I have done lots and lots of films over about 15 years with Tom Cruise, and I worked a lot with Daniel Craig, I have worked a lot with Angelina Jolie. Yes, there are a lot of actors that I have worked with over and over again, and I guess what happens is that you are in that kind of group and that's where you end up working.

What advice can you give to young people just starting out in their careers?

I think the best advice is to listen and learn, and that is why we have assistants and trainees on set.
The most important thing is that they get the opportunity to learn from others and although they might have gone to film school to learn the language and to learn the basics, they will learn much, much more on set. The most important thing is to remember that when they are starting on set, they are starting from the beginning. They need to try to be like a sponge to soak up everything in that they can learn.

On Christopher McQuarrie's *Mission Impossible 7* - 2020

Pascal Armant
France

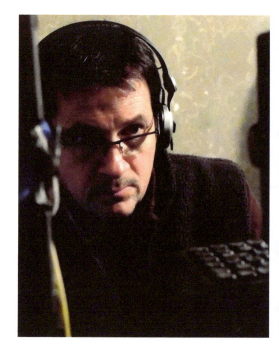

Pascal Armant

After musical and scientific studies, Pascal turned to location sound recording and has worked on numerous projects since 1985, first as boom operator and then, since 1990, as production sound mixer.

He has worked in every area of the audiovisual industry, between institutional, documentary, magazine and TV film, short movies and feature films.

Pascal Armant has been the sound engineer for several movies such as *Gauguin - Voyage de Tahiti*, *Intouchables* (César Awards nominee Best sound 2012) and *Le sens de la fête* (César Awards nominee Best sound 2018).

On shooting *Jusqu'à toi* directed by Jennifer Devoldère - Paris - 2008

What does it mean for you to be a production sound mixer? How did you enter the world of sound?

In my opinion, the production sound mixer is the first technical role that gives directors the possibility to transpose the dialogues and the stories they imagined writing the screenplay into sound.

I approached location sound through music: I attended an audiovisual school, the ESRA in Paris, with the idea of working in a recording studio, being a musician myself.

However, after doing some internships in film mix theatre and working on a few short films, I became increasingly fascinated by the idea and the possibility of working with sound for the cinema, as well as by the type of profession that requires you to always adapt to a new environment every day, in every corner of the world, and does not "lock you up" in a studio.

In a sound technician's career, how important is it to have a steady working relationship with the same director over the years?

Trust and loyalty are crucial in my work because it allows us to work in the best conditions, we know each other well and we know each other's preferences.

Often, directors have sound ideas in their head and they talk to me about them in pre-production or on set, and we start planning the things to record that will be used for future sound editing and mixing.

If some sets are noisy, the director may agree to change them or to adapt them following the sound engineer's indications, if there is trust.

How has this job changed, if it has, from when you started your career to today? Do you believe there is a "European" and an "American" approach to this profession?

Compared to when I started, this job has become more "technical", there is more material to manage.

When I started out, a Nagra, a microphone on a pole and a few other microphones were enough, but the results were not always excellent…

Nowadays, shooting is only done by wiring all the actors, giving wireless headphones to all the crew members who ask for them (director, editors, assistant directors, etc...) and adding a stereo (ambient) pair on the tracks we have left in our recorder.

My experience working on American films has given me the impression that European directors are more "attached", during editing, to wanting to keep as much of the direct sound as possible and therefore to be more attentive on set to the sound engineer's suggestions, such as redoing a scene for sound if necessary, waiting for the end of a noisy event like an airplane or a car.

Therefore, in Europe we often have the habit of intervening in the directors' choices and evaluations during shooting.

I think the skills and equipment of European and American sound engineers are very similar, but I think the Europeans have maintained a more "artisan" spirit, even in post-production.

With the Americans, everything seems to be very framed (timecode, electronic clappers, very precise numbering of the shots) with the aim of making the most of it and thus avoid wasting any time in post-production.

What is your main objective when you record the sound for a movie?

On set, my goal is to provide the director and editor with the best possible sound elements for the construction of the future soundtrack (dialogue, ambiences, effects, possibly music). I try to make sure that this can happen in the best possible "environment" with the other members of the crew (camera department, direction dept., etc.) as I do not like situations of conflict, this would affect my concentration and the quality of my work.

When recording, what makes you happy and satisfied and, on the contrary, what causes you to be disappointed or dissatisfied?

The films that I am most satisfied with are those where most direct sound has been kept, in accordance with the director's wishes, as well as films where we had to be innovative with the technical means used to "succeed" in a particularly difficult sound shot (for example in the film *Intouchables* - the paragliding sequence, or in the film *Le sens de la fête* - the musical sequences with orchestra).

On shooting *Les aventures de Spirou et Fantasio*
directed by Alex Coffre - Marocco - 2017

I also cherish the memory of some very powerful moments during filming, where the actors moved us particularly with their interpretation and where we had the impression of experiencing exceptional moments that belong only to the team present at those precise moments…

On other sets, on the contrary, I was sometimes disappointed not to find the spontaneity of the actors or the naturalness of certain live sounds, because they had been weakened/modified by special effects or because the music was too intrusive.

Tell us about your equipment and how you have organised your sound cart. Is there anything in your sound kit that you could never do without?

Today my gear consists of an AATON Cantar X3 recorder with the Cantarem, as a second recorder I kept the Cantar X2 and I have a Sound Devices 633 for the light units.

I use a set of Wisycom and Audio-Ltd A10 for the radio booms, all in a Sixpack or Octopack rack with Betso antennas.

The transmitters are chosen according to the costumes and actors to be miked.

I have an Audio-Ltd mini-TX, a Sennheiser 5212 and a Wisycom Mtp 40.

As capsules, I alternate between Sanken cos11, DPA 4060 and Countryman B6.

I often use the Sanken CUB01 for camera cars or as plant mics.

At the moment, my favourite microphone for the boom is the Schoeps Cmit 5U or the Mini-Cmit with Cinela suspensions and windscreens, but in certain situations I use the Neumann KMR 81 and the KM 150.

For ambience, an LCR Sennheiser 8000 series or double MS Schoeps CMC.

As for the Comtek system, I use the Sennheiser G3 series.

For headphones, I use Sennheiser HD25 and the poles are VDB and Ambient.

As monitor, a BlackMagic dual screen.

When you read a script, what is your approach towards the sound? Which types of scenes worry you the most? What is your next step after reading the script?

When I read a screenplay for the first time, I don't immediately analyse it in technical terms but first enjoy the story.

Then, I proceed to make a precise perusal in which I take note of all the complicated aspects such as the number of actors per sequence, scenes in cars or moving vehicles, scenes with music, playback, etc…

After the reading, I prepare a list of questions to ask the director (set-up of the shoot, how many cameras, etc…) and try to attend a script reading meeting with the other department heads (production designer, director of photography, costume designer, director) to prevent any difficulties.

I try to always be present during the location scouting in as many locations as possible to give my opinion and to assess how they are for sound.

The sequences I am most concerned about are those containing many actors and multi-camera shots.

Have there been times in your career when you could not secure a usable sound – and how did you react when there are unsurmountable difficulties? Any interesting anecdotes you would like to share?

Of course, there have been times in my career when I have not been able to overcome certain difficulties and where it has not been possible to keep the direct sound, for example, when actors speak too quietly in a very noisy set like on a seashore, in a vehicle, near a highway or on an airport runway.

Today, movies are increasingly shot using multiple cameras, which makes operating boom extremely difficult. Could this profession have survived without wireless microphones?

Absolutely not, you couldn't work nowadays without wireless mics.

How many people make up your team? Have you been working with the same boom operators for many years? In your opinion, what makes for a good boom operator/assistant?

On a major film, my ideal team consists of myself, two boom operators and a trainee. For certain music sequences, one more person may be needed.

On shooting *Be You* directed by Maïmouna Doucouré – 2022

I would say that for a standard film, three people may be sufficient.

I have been working with the same boom operator for over twenty years, we know each other well and it simplifies the working relationship.

A good boom operator must have good physical stamina, they must be brilliant, quick, they must be able to adapt when the actors improvise or when the directors make changes to the scene at the last minute.

A good boom operator must be familiar with their equipment, know how to microphone the actors, and above all have a close relationship with the rest of the crew and always be an ally (camera dept, direction dept, costumes, props, etc.).

How important is it to interact with the post-production team before, during and after shooting? Do you agree that post-production can significantly improve your work, as well as make it worse? In your movies, do you always manage to follow up the post-production work?

It is important to be able to talk to the post-production team before shooting, to accommodate their preferences if that's possible.

Sometimes we start shooting without the post-production team being fully staffed yet, then I act according to my personal taste and habit, but it is riskier.

I do agree that post-production can easily improve our work or make it worse, although the latter is rare in my opinion.

Depending on how busy I am, I try to go through sound editing and mixing at least once.

On set we might give our very best to record any small nuance of sound, but then movies are mostly watched on a computer or on TV. In your opinion, how attentive is the average viewer to the sound, or rather, how much do they really understand the quality of good sound?

In my opinion, even though many viewers watch movies on small screens like cell phones or tablets, they are still quite receptive to the quality of sound we provide them (in fact, they often listen with headphones or earbuds, so they can perceive certain subtleties of editing or mixing).

On shooting *The smell of us* directed by Larry Clarke - Paris 2014

95

Stefano Campus
Italy

98

Stefano Campus

Since 1995, Stefano Campus has been working in the field of professional audio, first as a live sound engineer, monitoring concerts and live amplification, then he worked on music mastering and room acoustics, providing corrections and sound measurements. In 1999, he graduated in Sound Technique from the Centro Sperimentale di Cinematografia and has been production sound mixer for numerous films for the cinema, including: *Fortress* (2021) by J. Woodworth, *The Story of My Wife* (2019) by I. Enyedi, *Ordinary Happiness* (2018) by D. Luchetti; *Lucia's Grace* (2017) and *The Complexity of Happiness* (2014) by Gianni Zanasi; *I Cassamortari* (2020) and *Il Permesso - 48 ore fuori* (2016) by C. Amendola; *Saimir* (2003), *The Rest of the Night* (2008) and *Black Souls* (2014) by F. Munzi, with which he won the David di Donatello; *My Class* (2013) by D. Gaglianone; *The First on the List* (2011) and *Feather* (2015) by R. Johnson; *Horses* (2011) by M. Rho; *News from the Excavations* (2010) by E. Greco; *Sonetaula* (2007) by S. Mereu; *Fascisti su marte* (2006) by C. Guzzanti; *Caffè* (2016) and *Red Like the Sky* (2004) by C. Bortone; *Palabras* (2002) and *West* (2000) by C. Salani; *Round the Moons Between Earth and Sea* (1998) by G.M. Gaudino. He also alternates between travel and extensive work assignments abroad including Australia, Spain, South America and the USA. In 2009 he taught Sound at the European Film College in Ebeltoft, Denmark. Since 2011, he has been a lecturer at the Centro Sperimentale di Cinematografia (National school of Cinema, in Rome).

Ordinary Happiness directed by Daniele Luchetti (photo by Paolo Ciriello) - 2019

You started your career in the "world" of sound, working with post-production, live music, etc. How useful and important was it to have this knowledge and experience in the way you work today as a production sound mixer?

My father had an audio service company and I used to spend my summer vacations with him, working alongside him during live concerts, I had a lot of fun.

Before enrolling at the *Centro Sperimentale di Cinematografia (Italian National film school-aka CSC)*, I had worked in post-production, but during my studies I realised that I enjoyed more working with live sound on set rather than in front of a computer. This wealth of knowledge and experience has been very useful to me in the specific field of film sound recording.

In the world of music, clean sound, high fidelity, and technical phenomena, in absolute terms: these are the priorities. With sound for films, having a technically clean recording and the best microphone are not the only priorities: narration and storytelling are paramount.

What I like about location sound is finding that combination, that moment that can bring an unrepeatable performance to life.

For example, in the live recordings of the great Peter Gabriel, imperfections are evident, but the emotions of the singer, the drummer, or the bass player give that extra something, that error that is full of humanity.

The same goes for location sound: we often find ourselves dealing with and solving difficult situations. I think that recording sound for cinema is not only about recording dialogue, but the signals also that help us understand the script. It is also about providing other crucial sound elements that make up the entire film. The fact that I have also gained experience in music production and film post-production gives me a clearer vision of what the result will be like.

This experience, which was especially useful at the beginning of my career, has now become an acquired instinct; for example, I can easily sense which off-screen sounds or wild tracks are essential.

Edgar
So, in a way, knowing how post-production works helps you to know what exactly you may need to record at that moment on set.

I am often too greedy: I like to record everything that will be useful in post-production. Obviously, it is always a matter of priority, of importance.

The sound can also be worked on later; if there is a scene where the light is "straddling" (changing), you have to act quickly by focusing on what is happening.

Edgar

Has your experience in music helped you to build up your own personal taste, has it helped to "stimulate/fine tune" your ear more?

It was a real training, also because the sound is in a particular setting; for example, you easily get used to the continuity of sound, compared to an impulsive sound, and it is very important how the sound will make its entrance.

There are many ways to record music. Let me give you an extreme example: an orchestra of 160 elements can be recorded with 160 microphones, each one close to each source, capturing only the direct sound and then building the ambience later.

I prefer other techniques where the microphones are further apart, and I try to use room acoustics for added value. Having these two extremes in mind means that every listening session you do forces you make different choices. I like environmental interaction because, in addition to transferring information related to the dialogue, I can, for example, describe a place, underline the passage of time, dilate or contract the perception of a sequence or suggest an emotion. Often, simply moving the microphone away from the source is enough to increase communication efficiency.

Drawing on my knowledge of recording music, I don't hide that I am tied to the Decca Tree set-up that contains a lot of ambience. My taste is linked to a very precise concept of sound spatiality. Today I notice a bit of flattening of the sound fields. The tendency is to always go to the foreground, on the absence of fields, on the use of sound rather than on the narrative sound. This background is part of me: if, for example, I am asked to record the most direct sound possible to be then processed and treated later, in some way I experience it as a limitation.

How did you end up working as a production sound mixer and what were the beginnings of your career like?

I have always been fascinated by sound in all its forms, from music to cinema. I'd done technical studies in engineering, but I was in trouble because the artistic and creative part was missing, and I didn't feel comfortable. Then, by chance, I heard about a training course on Pro Tools software.
I found it very interesting, and I was asked to teach a practical module at the next course, in 1995. At the time, Pro Tools was used only and exclusively in music. I delved into this software, deepening my knowledge and understanding of it, which opened the door to various work experiences in recording studios.
In 1996, I was interested in the Sound Technique course at the *Centro Sperimentale di Cinematografia* in Rome, so I sent the audio materials from my previous work experiences for selection, and I was admitted. My career followed a natural path after that, and at the end of my first year I was already gaining some work experience in my free time. During my years at the *Centro Sperimentale*, my interest shifted from post-production to location sound.
I remember an important experience with my production sound teacher Bruno Pupparo, who asked me to give him a hand with the environmental recordings of a film he was shooting in Salento, called *Sangue vivo* (2000).
I can only say that, at that moment, there wasn't much time to think, we were only six students every three years leaving the *CSC* and we were immediately thrown into the job market.
There were no internships at the time.
During our studies, I remember that my classmate Alberto Amato and I did some outside work. We shot a short film with a "serious" crew. There was a track with a dolly on it that creaked, made noise and there was also a parquet floor. We were thinking about how we could solve this problem.
During our lunch break, everyone went off the set except us and we set about doing something that would be unimaginable now: we removed all the wooden wedges that were put in and repositioned them so that the track would no longer make noise.
You can imagine the reaction of the grips with thirty years of experience when they saw what we had done.
Books did not teach you how to deal with other professionals.

So, at the beginning of my career, I might get into an argument with the electricians because I was used to recording music in the studio, and it was unthinkable to cut a voice below 60 Hertz.

Also, I thought that a generator on set could only be well placed if ten kilometres away.

The great advantage of doing an internship is to see this reality alongside experienced sound engineers. They can give you a better understanding of the set, which is a very particular environment where there are many priorities beyond a clean recording or pure technique. It is a job that relies on collaboration; you work as a team and relationships are crucial.

At the end of my last year of school in 1999, I made my first film as a sound mixer, *West* (2000) by Corso Salani. I was 21 years old, and I felt a great responsibility.

Every evening I listened to the recordings through headphones with my eyes closed, obsessed by these sounds.

Even finding the equipment was more of an adventure than today, so the difficulties were those of navigating a context that I knew little about.

104

Edgar

After school did you start immediately to work as a sound mixer or was it your intention to be a boom operator?

I did lots of work as a boom operator, but they were small jobs, documentaries, short films, commercials where I did everything by myself, both as a sound mixer and boom operator. However, besides some practical exercises during my studies at *CSC*, I've never worked as a boom operator for another sound engineer, and I lack that [experience].

I loved the dynamism of working as a boom operator, but I found myself quite overwhelmed, since I had many offers both as a sound editor and as a sound mixer, and often I did both also for the same film.

Today, I think that good performance requires the use of various professionals, so if we do our best during filming, another professional should take care of post-production with the proper distance and freshness.

I've learned all these things at my own expense, and I try to pass this on when I teach my sound course at the *Centro Sperimentale*. Here we actually have a rule that of all the students doing both location sound and post-production, those

Horses directed by Michele Rho (photo by Daniele Mantione) - 2011

who are production sound mixers in a short film can only supervise in post-production, without having an active role; they can perhaps work on the post-production of another short film where they were not involved in the shoot, so that they alternate.

Do you think that creating a good network of contacts and knowing how to move in the sector is as important as being good at the job on the set?

Professionalism and competence are essential on set. Entering the project, working well with other people, choosing the right equipment, analysing the script, talking with other departments: these are crucial requirements of our work as sound technicians.

Furthermore, there is another important skill which is the ability to manage good relationships.

Let me also add that it is essential to not chase your ideal of perfect sound but rather to listen to the person who has conceived the film. We have to be at the service of the directors' creativity, and if we can work within this realm then there is a lot of room for creativity and many opportunities to express ourselves.

The best compliment for me is when, during the first editing sessions, the director tells me: "I have to thank you because you have given me a lot of good material, a lot of possibilities and so we are in the best possible position to choose".

Furthermore, technicians like us are always "caught in the middle" between the director who wants the maximum and production, which by definition has to set limits and make ends meet.

I'll give you an example: if I am making a 500,000 euro independent film (low budget) and we shoot in the streets, I have to imagine that there would not be any possibility for street closures around the Pantheon for a radius of 3 km, since from a production point of view it will be unthinkable.

I recently happened to have a wonderful experience on a short film by a friend of mine who is a director, shot in Sardinia with a very low budget, where we had an electrical generator which was never silent.

We managed somehow, with a very helpful troupe, and this is the right attitude.

And now the basic question that I ask myself is: "How can I help you"?

What is your main goal when you record sound for a movie? Which characteristics should the sound that you record have to make you satisfied and happy and, on the contrary, what can make you unsatisfied about the sound you are recording?

Today I can tell you that there is no absolute value.

The most important thing is to stay focused on the film, on the story, understand the film that you are doing, examine the script in-depth, participate in the discussions with the direction department.

Right now, I am preparing for a beautiful film that I'm going to shoot at the end of August (2021), one of those films that rarely happens, with an American/Belgian director who involves me in everything. She lives in Brussels; she sends me emails and we exchange many ideas, and for me this means getting into the film. In order to find solutions, we need to understand how the director shoots the film, which style they have, what they will use: are there long takes? Are there handheld cameras? Dolly? Steadicam? Which is the right setting for the film? For me, it is important to render naturally what my sensibility understands from the script. Not if a line comes out wrong, or if we need to do it as wild line. I feel sorry when we are not able to recreate the right setting, especially the auditory one, for a specific era or context.

For example, in a film shot recently in Budapest titled *The Story of My Wife* (2019), we were supposed to recreate 1920s Paris: we had lots of extras, beautiful costumes, vintage cars, but offscreen, on the other side of the camera, there was modern-age traffic, impossible to stop. Therefore, the sound of that era was created entirely in post-production.

Edgar
So, to summarise, we can say that your main goal when you record the sound of a movie is to understand, to enter into it and be at the service of the film itself. Basically, you don't set other additional "elementary" goals, more related to location sound itself.

Everything is a consequence. In an ideal situation, I would like to provide three types of microphone solutions and so I have always the foreground with the lavaliers, then the natural field of view of the camera with the boom, and when there is an interesting environment, I add a stereo mic or a quadraphonic setup, far from the action, only for reverb.

I like to give this richness to be able to reinterpret from a sound point of view the sequence even in post-production. With these three listening points, the sound field can be controlled dynamically, and if, for example, there is too much contrast between the wide and the close up, we are always able to correct it, and therefore to create a sound perspective independent from the framing.

In *The Story of My Wife*, during an intimate scene shot with two cameras and different lenses, the husband and the wife were whispering: the sound of the lavaliers wasn't very good, and so in order to obtain a good sound with the boom, we asked not to shoot with the two cameras simultaneously. There was a strong spirit of cooperation, and above all the technical sensibility of the director allowed us to obtain an exceptional result.

Anyway, it is necessary to have a good dose of flexibility to understand that some things, if shot simultaneously with two cameras, are better. For example, I saw Luca Bigazzi and Luan Amelio (infallible camera operators) at work, shooting simultaneously with more cameras dancing among shots and reverse shots. This open-mindedness really helps, we have to interact with the troupe, forgo chasing the perfect recording, in an absolute sense, instead we need to evaluate situation by situation.

Edgar

So, we can say that you are happy and satisfied with the sound you produce when you believe that the sound is good for that type of scene and director.

Absolutely yes.

The objective is to understand exactly which kind of sound to create for that particular situation, so that we can face the creation of the sound of the film properly. Usually, from a technical standpoint, the greater the difficulty the more motivated I feel. For example, on the set of the movie *My Class* (2013) by Daniele Gaglianone, I was intrigued by the objective difficulty of the situation. We were in a real classroom; the only actor was Valerio Mastandrea interpreting the Italian professor with twenty foreign students. There wasn't any script, it was pure improvisation with two cameras and with action lasting one hour. The solution was to place a lavalier on everyone and record on 24 different tracks to have total freedom.

Let's talk about the equipment that you use and how you organise your sound cart. Do you change frequently your equipment? Is there anything in your kit you can't do without, and which piece of equipment follows you in all your films?

From an emotional point of view, in my case the most precious tool is a set of Leatherman multi-purpose pliers.

It was a gift from a friend of mine for my very first film and I am very attached to it. I can almost say I never leave the house without that tool in my belt.

Together with Sandro Ivessich Host, the boom operator I've been working with for more than ten years, we change our equipment very often, and we enjoy experimenting with new things. The double kit was a real game changer. I remember very vividly how we worked before having that: you had everything ready on your cart, then suddenly they said: "Now the Steadicam" or "Now is camera car" and you had to disassemble everything and reassemble the entire cart, and then jump in the boot of a car.

With the double kit, we are always ready: we have two identical set-ups, one on the cart and the other on the shoulder bag with two Aaton recorders and double radio receivers set exactly on the same frequency. In a few seconds, we are ready to face sudden changes and we can also afford to have a moment of reflection on practical problems, such as how to better secure the safety belt so that it doesn't scrape on the mic or finding the perfect place for a plant mic. Certainly, it's an expensive choice, but having twice the equipment enables us to optimise every situation.

Even the second assistant is fundamental.

A production sound mixer works concentrated and records constantly, the boom operator, on the other hand, never has any spare time, with all the actors he has to mike and try the scene with the boom. So, if there is a second assistant who can manage headphones, batteries, soundproofing, etc., then it's possible to work more serenely and be more aware.

The time when there was the Nagra, a cable, and the MKH 416 is over - and it's never coming back. Then, if we want to have an atypical experience and redo a film in that way, so be it! I would love to rediscover acoustic fields. Today, though, we work differently in feature films, and I believe it's the right thing to have a sound team consisting of three people. And we've had this philosophy at the *CSC* institute for years.

Il *Permesso-48 ore fuori* directed by Claudio Amendola
(photo by Giulia Bertini) - 2017

Another fundamental piece of equipment is the digital wireless stereo set up. This is very useful on set, where the schedule is always very tight; it's truly a blessing to have a stereo mic connected to a single plug-on, which also makes a digital conversion and brings two embedded channels to the receiver without losing its quality and without using the cable.

Having this opportunity allows me to quickly find the perfect placement to record a murmur, an interesting scenario or screams in a tunnel, or in a church.

As for the wireless system, we use Lectrosonics and Zaxcom. And as far as other equipment is concerned, we are traditionalists: we love Schoeps microphones, we have all the series Colette and Cmit.

We don't disdain the Neumanns either, and we also like the Sennheisers; I would say that, more or less, we have them all. As lavaliers we use the Sanken Cos 11, I've also tried other models but knowing them very well, combined with the analogic sound of the 411 Lectrosonics ones and Schoeps, I can obtain great uniformity.

Regarding the boom, if there's time, I use cables. Especially in those scenes where there is a lot of dynamics, I enjoy rediscovering classic simplicity as we often did in *The Story of My Wife* with a cardioid microphone.

But sometimes you must "wage war" and it is an incomparable luxury to have the Zaxcom digital wireless system with the ability to control the input sensitivity of the transmitter from the sound cart through ZaxNet, without disturbing the actors. I have to say that the Zaxcom system interests me greatly, precisely for this ability to remotely control the transmitters, besides the dynamics of the converters and the possibility of having a constant backup on the transmitter microcard in case there is radio interference.

Edgar
Which headphones do you use?

I'm comfortable with the Sennheiser HD 26, above all for their lightness and compact size, very classic but with this particular function of the limiter (ActiveGard), so if there is an unexpected loud noise, the headphones automatically attenuate it and you don't have to lower the volume to protect your hearing.

When you receive the script of the film you will be working on, what do you do? How do you organise your work? Are there some types of scenes that might concern you more than others?

The sooner I get the script, the better it is for me.
As a cinema lover, I do the first reading like a book, and I see what effect it has on me from an emotional point of view.
Sometimes I can imagine the film, other times I don't really understand what kind of film it is and so it becomes interesting to go deeper into it with the director.
I really love when the director takes a morning or an afternoon off to drink a coffee together and explain to me what kind of film he or she wants to shoot. If the director does such a thing, I will probably end up accepting the assignment, even if the pay isn't excellent. After that I analyse the script scene by scene: how many actors there are, if there is live music, possible criticalities, sync sound effect, etc…
So, I try to arrange the work both from a technical and organizational point of view.

Another fundamental aspect is the difference between having the opportunity to do a recce and when this is not possible.

Being aware of a problem in advance gives me time to prepare, the fact of seeing previously some of the selected locations for me represents a benefit since I know what the criticalities are, and I can come up with an idea to solve them.

In the film by director Luchetti titled *Ordinary Happiness* (2018), for example, we had a location in one of the most congested streets of Palermo.

We had plexiglas panels installed on the windows, which reduced considerably the background noise.

Another step to take is always talking with other departments. For instance, during a film that I shot at the beginning of my career, titled *Red Like the Sky* (2004), I worked with a great production designer, Davide Bassan, who made a lot of beautiful films. We had a problem with the main location, which should have recreated normal classrooms, but actually had 20-meter-high ceilings.

There was a natural reverberation unsuitable for those shots. We consulted with each other, I did some surveys and, having both a director and a producer listening to me, we agreed in choosing sound-absorbing materials as furnishings, like carpets and tapestry.

Furthermore, grips helped me create a small invention (I still get teased about it to this day), which we named "*armadilli*". They were classic cinema panels in polystyrene two meters by one, to which we attached sound-absorbing materials, ashlar on one part and pyramidal on the other, double layer, which we placed on some stands: this helped me to interrupt the reverberation many times and to completely change the acoustics of some locations.

Edgar

Are there some scenes that are more worrying than others?

Sure, for example when live music and dialogue must coexist in a long take, I am very careful. They are some types of scenes where one needs to apply certain criteria, above all if you want to give maximum freedom to the sound editing.

Those scenes represent more of a challenge rather than a concern. It's good when you have three booms, ten wireless microphones, a playback to be sent via headphones, and even a microphone on the scene with an actress singing and an entire band to shoot live.

It's very satisfying, especially when we can share the level of difficulty with production and they give you the necessary time to implement everything.

Some directors are more sensitive to the sound aspect, others less so. Does your approach to your work change according to the director you have to work with? Can you provide some examples?

There are projects and projects, depending not only on a productive point of view, but also on the creative one, the director, the style, and the poetics. The guideline is not the technique. Technique and equipment are functional tools for the fulfilment of ideas. I'll give you two examples.

The first time that I met director Gianni Zanasi we exchanged a few words to understand what kind of film he wanted like to make, and he asked me immediately: "But do you have wireless microphones? Because I love the sound of the lines here, upfront".
As a technician, I didn't need further information to understand which idea of sound he had in mind.
During our collaboration, his idea changed and in the second film we did together, he didn't talk only about dialogues in the foreground, but also about the ambient sounds that had major importance, since he wanted a more organic sound.

The opposite example is director Francesco Munzi, who I've known since our time at the *CSC*, and I worked on all of his films. When we were preparing *Black Souls* (2014), (he kept me constantly updated on all those preparation developments) he said: "We're going to shoot in Aspromonte, a really peculiar place, so I would like this place to be treated as a real character".
At that point, we organised with a second unit to record ambient sounds of all the environments we visited in quadrophonic.

What is your intention, what do you aim to record on set so that the scene is "covered" from a location sound point of view? What do you try to provide to post-production?

First of all, I reaffirm the importance of providing everything that can be necessary to post-production to give the freedom of construction and ambience setting, varying the connection between direct sound and interaction with the environment.

Therefore, even in those situations where a single boom seems enough, I always put a radiomic too, because for me it's important to have a foreground (lavalier), the natural field of the boom (relating to the framing field), and another figure (stereo), which is farther in order to obtain the extreme opposite, which is the maximum interaction with the environment.

It is necessary that what we see at the movie theatre doesn't only "seem" true, but it has to be true, trying to record even what is outside the screen, offscreen.

If, from the visual point of view, we are anchored to the frame, from the sound point of view we have a full 360 degrees to use.

Characters (actors, extras) don't only exist when they have to enter the scene.

Obviously, this can also be done in post-production, but before leaving the set I like to collect as much authentic material as possible, like a murmur in a foreign language.

Some years ago, at the central market in Marrakech, we recorded typical environments in their uniqueness, with snake charmers playing, roaming animals, lots of "walla" in the original language which helped us to create a full and authentic sound stratification.

I always enjoy recording the truth of a specific moment, even if Foley artists always have everything in their archive, also because the final result is the mix of various sounds, real and recreated elements. Many times, the sound utility deals with recording something interesting offscreen.

Going back to a question you asked at the beginning regarding my experience in post-production, I can add that on the set of *Horses* (2011), a film set at the end of the 1800s, while shooting a scene in a medieval village in Colle Val D'Elsa (Tuscany), the director became interested in sound shooting. We were shooting an establishing shot in a square that come alive early in the morning. After many takes, I told the director: "Listen, since this square has its own particular reverberation, why don't we have that lady washing the floor over there dump all the dirty water she has in her bucket?"

The director understood what I was talking about and therefore we added this small detail, which characterised the sound of the square even more realistically.

The story of my wife directed by Ildikó Enyedi (photo by Hanna Csata) - 2021

I know the sound of digital reverberations really well, even the most sophisticated ones, and I think that well-recorded reality is still unbeatable.

If a place has peculiar acoustics, with the proper solicitation, what could be better than recording it in quadrophonic, in 5.1 or with a Schoeps ORTF-3D?
Providing this wealth to post-production is an added value; then it will be up to the sound editor to listen, select and evaluate if this material is useful or not.
I believe that the goal must be to create an indissoluble sound mix of all the elements that make up the sound of a film: dialogues, backgrounds, effects, and music.
We have no choice in such instances when productions, attracted by an "All Inclusive" under-priced package, forgoes using double booms, radio mics, stereo, wild tracks, effects, murmurs, etc. You can be sure that at the premiere of the film you will end up with the sound from wireless microphones alone, cut word to word and recreated totally aseptic backgrounds.

Unfortunately, in the Italian cinema industry, this is a popular trend. As cinema and sound professionals, and lovers, we must protect a certain type of workflow.
Besides the surveys, we need to start thinking about the sound that will be useful for that particular film, according to the budget, purpose, and target audience: is it a TV film or is it for the big screen? Is it an auteur film or is it a commercial film? And above all, it is crucial to work as a team with post-production.

Edgar
Can you also cover an offscreen dialogue with a second boom?
Or at least are you inclined to do it as an intention? Is it useful?

I actually changed my mind after working with Léa Seydoux. She didn't want to have a lavalier except in the shots where she was in frame, nor did she want her "listening shots" to be recorded with a second boom.
After a few days of seeing her in action, being an incredible actress, we understood that we could have edited only her close ups or her shots, because it was very emotionally intense.

In conclusion, it is essential to have the experience and sensitivity to understand if the off-screen actors are really acting or just "serving" to help the framed actor.

The second boom operator is essential, especially in those contexts where I need many options: there can be situations within a scene in which the dynamics are very wide, so you can go from screaming to whispering. In that case, I need a shotgun microphone on the whisper because I need to be very close and with a cardioid microphone on the scream.
I leave the boom operator free, not just horizontally for the focus, but also vertically to adjust to the dynamics of the scene, to make sure that the boom records the dialogue as naturally as possible.

Have you ever encountered some very difficult scenes when you did not have solutions to obtain a usable sound? How do you behave in those cases? Do you have some examples to share?

The most difficult situations are those in which wireless microphones don't have reception, and today wireless mics are vital. The film *The Story of My Wife* happened to be shot in the bowels of an all-iron ship, where even the best antennas don't work well, and so it was necessary to be as close as possible to the actor.
In this kind of situation, having the "double setting" was of vital importance: I took my portable kit which allowed me to quickly solve the problem, even if that meant renouncing the comfort of the cart.
Being seated far from the set, focused on your monitor in the calm of the video village, makes it possible to listen carefully and make the best choices.
However, in other situations as a matter of coverage, of distance, you have to understand immediately if there is a problem and organise consequently.

Edgar
Many times we don't have control and that is frustrating. How do you handle this condition? Do you accept it and live happily, or does it distress you?

I went through many phases; therefore, having experience is fundamental.

When I find myself faced with problems, I ask myself the right question: "direction, actor, producers, how can I help you?"

If there is no solution, it's important to expose the reason clearly.

In a movie set in 1920s Paris but re-enacted in today's Budapest, even the producers apologised for not having been able to put us in an optimal condition.

If there is this kind of approach, you can accept some situations since you know that, anyway, the sound of the film will be well developed.

Instead, when you are faced with unsuitable situations, where you are asked to do impossible things, you get over it.

Usually, I'm glad to give a set of headphones to the costume department, interior designer, or generator operator and to other departments to make them step into our world, where sometimes it is difficult to explain with words what our difficulties are.

Other times they warn me: "Look, today we have a very noisy location", and instead we are in a situation with a lot of background noise, yes, but actors speak at a proper level, the signal is good, lines are in focus, so everything it's ok; for me it's a relaxing day because the background noise will make everything uniform.

The real problem, instead, is when I have two characters murmuring in a room, you can't put a lavalier on them because they are half-naked, sheets make noise and all the troupe is behind the camera, generator, dimmer, ballast… but it's important to isolate the silence since it is an integral part of the emotion of the scene.

What do you think should be the proper profile of a sound engineer and a boom operator regarding the other departments? How do you and your team connect with other departments such as direction, camera, costume, actors…? Can you give us some examples of an effective collaboration with another department, which helped you to achieve the desired sound? And on the contrary, do you have any anecdotes when this collaboration failed and made things difficult?

First thing is to have fun and help others because cinema is teamwork.

I'll give an example: we were miking Claudio Amendola (director and actor), I asked Sandro (boom operator) to get the capsule out of the button and Claudio with the same mentality said: "Why don't we patent buttons with a built-in microphone?".

There! When you have this kind of collaborations things work. I understand the camera department's critical moments, when the light is changing, I do my best to not let them miss out on the moment, or when in a period film you handle costumes that are unique pieces, it doesn't make sense to put gaffer tape or other things that could damage them, and risking ruining the relationship with the department as well.

We are all working together to produce a movie, it is very much a team effort.

I happened to record a dialogue between two workers in an iron and steel factory where the machines couldn't be turned off, the common thought was that the scene should be dubbed.

Obviously, I couldn't stay too close with the boom because they would have shouted, but with an unorthodox solution, that is using a dynamic and less sensitive microphone, we achieved our goal.

Usually, finding the right balance between sound layers helps the actor in creating a genuine performance.

Basically, I love cinema, so if a scene results in being dirty but beautiful at the same time, I don't feel in trouble.

Edgar

Do you have also any negative experiences, or rather, examples of non-cooperation?

Recently, I happened to shoot an indoor scene in the Parioli neighbourhood (Rome), where at some point the actors were moving to the terrace while conversing. I arrived earlier on the set, and I saw the generator placed in the street right next to the window where the action was taking place, 3 or 4 metres as the crow flies.

I said to the production manager: "I don't want to put the cart before the horse, but the power unit is too close to the set". The answer was: "I know, but there wasn't any other option. Please, you write down that this is not good, and then post-production will think about it".

I felt strongly against that hurried answer. Therefore, feeling a responsibility towards the director and realising that it would've been easy just to move the generator away, I did something which could seem paradoxical.

With calm, I said: "You know what, I don't intend to write, I'm here to do sound for a film, if it's not possible today, I'm gonna leave, you can call someone else" and so I did.

I explained the situation to Sandro (the boom operator), and I went to the café to have breakfast.

30 minutes later the production manager called me back saying: "Sorry, we started off on the wrong foot, the generator has been moved at a fair distance", then I returned to the set.

For sure that production manager won't call me again, but if there is sloppiness and laziness, I am not doing this.

There's always a fine line between those who are good but accept everything and those who cannot compromise in some situations.

What is your thought about wireless microphones, the use of the wireless boom, and shooting with two or more cameras? Do you place lavaliers in person or is it a task for your boom operator?

It depends on the situation.

Regarding the wireless boom, I have the example of *The Story of My Wife*, where we went back to the purity and simplicity of the cable connected to a Schoeps cardioid.

There are situations in which you could have incredible dynamics and, despite having at my disposal the digital Zaxcom, if there is time and there are the proper conditions, I still prefer to use the cable.

Sometimes radio microphones are fundamental.

All technical equipment can turn into opportunities, if used wisely and not with "autopilot".

So, it's clear that there are objective problems compared to twenty years ago when it was easier to run a cable on the set.

Some years ago on a set, we still did not have the digital plug-on; there was an actor who was screaming so much that I preferred to approach the boom operator and go with cable, rather than stressing out the compander of the analog plug-on.

The boom operator (Sandro Ivessich) places the lavaliers, but we always have a big debate concerning their optimisation. In some cases, if there are many actors to be miked, it can happen that I give him a hand, but he always makes the final check.

Placing the lavaliers is his task and this is totally fair, I am sure that he will do it better than me and he is very good at coming up with solutions and keeping good relationships with the actors.

When we are away shooting, in the evenings at dinner we discuss, for example, where is better to hide a lavalier if there are particular costumes or situations.

You are the artistic director of the Sound course at the Centro Sperimentale di Cinematografia in Rome, one of the most important schools of Cinema in the world. What do you try to instil in your students and what do you recommend to them when they finish school, and they are about to enter the job market?

Before anything else, there is the love for cinema, and it represents the basis on which everything else should be built.

CSC is first of all a school of Cinema and therefore, even during the selection rounds, the interest in cinema represents a significant decisive factor for prospective students.

I try to communicate the value of collaboration; the perfect recording does not exist, in an absolute sense, but there is the dialogue with the other departments, something that students have the opportunity to experience, since the CSC encloses all the departments, and it represents a sort of test for the real world of cinema.

In conclusion, don't be entrenched in your position and think only about what is good for your department, but try to help with generosity to make a good film.

I have to monitor their growth somehow, reminding them that there is an appropriate time to learn this profession. Practice done at school, graduation films with their peers and all the other experiences programmed in their study plan are important. I strongly recommend them not to jump immediately into the job market.

In my opinion, the best experiences started at Centro Sperimentale di Cinematografia, from the relationship with my colleagues of the directing course (Munzi, Mereu), editing (Benevento), photography (Radovich) with whom I found myself doing the first films after school.

What I want to highlight as a responsible professor, is not to limit excessively the field of study to one's specialisation. For this reason, my students can specialise only during the last year, so that they can experience both shooting and post-production during the first two years.

At school, when we do a practice film of a short movie, we

organise the whole team already during the preparatory phase, not only the production sound mixer and the two boom operators, but also who will edit the dialogue, the background sounds, effects, music, and who will do the mixing. This way, during the entire experience we can have meaningful dialogue on every essential concept.

We proceed harmoniously, with the hope that outside of school too we can work with this same principle.

I find it absurd that in 2021 there can often be such a clear distinction within the same department and that there is a lack of proper balance between the work done on set and the work done in post-production.

It's not easy to reconcile the job on the set and the one as a responsible professor, but I believe that it is important to educate generations of new sound engineers to have this wider vision, and to be well versed in post-production too.

During the production of the film *The Story of My Wife* there was this kind of collaboration.

Post-production was done in Germany by an all-German team, but during the surveys I met the sound supervisor, and we talked a lot about the sound setting of the film.

For example, he asked me to record specific sounds at the Museum of Historic Ships in Hamburg that he wouldn't easily find somewhere else.

Furthermore, in one of the locations there were many loud noises. I was worried and I sent over the recorded material to receive a quick feedback.

He replied by sending me some tests on the material processed with RX (software for noise-reduction) to let me hear the result, reassuring me that the problem could be easily solved. And so I continued to do my job serenely, knowing that I had another reliable set of ears listening. Organising the work in this way even at school, for me, is very exciting.

Edgar

Coming back to students finishing their course and leaving the school, what would you recommend afterwards? Shadowing a sound engineer as a mentor? Gaining individual experience?

Internships are fundamental because they represent a gradual entry into this profession, with adequate responsibilities for that specific experience level. This way, the recently graduated student can understand how the world of cinema works, without skipping steps.

I encourage students to be patient with their professional growth because it is important to protect their enthusiasm: you will not enjoy making a film if you don't have the right experience to deal with it. This could actually damage that initial positive mood and freshness that should be preserved as long as possible. And I say this because I went through it.

Since '99, the year of my graduation, I worked non-stop for three years, without still having gained the right experience. This fact caused me great distress in facing complex working situations, there was too much pressure and the resulting fear of failure.

It is necessary to enter this world gradually, with the proper help to succeed in managing even the most difficult situations.

Horses directed by Michele Rho (photo by Daniele Mantione) - 2011

Nakul Kamte
India

126

Nakul Kamte

Nakul Kamte is one of the most influential and award-winning sound designers in the Indian Film Industry.

Based in Mumbai and owner of Hearing Binaural, he boasts a wealth of experience in recording, editing and sound mixing for motion pictures and film, and he is primarily famous as a pioneering "live recordist", capturing sounds around the world. Definitively Nakul Kamte is credited with having greatly helped the Indian Film Industry (which initially preferred to have dubbing over live sound recording) to realise the importance of sound and therefore to prefer dialogue and ambient sound recorded on location.

128

Nakul Kamte on set of *Lion* directed by Garth Davis - 2015

With your work, you have contributed significantly to the transition between the "dubbing era" and "on-location sound" in the Indian film industry.
What can you tell us about this change? Has it been difficult to move decisively toward recording sound on set in India?

Yes, it's been extremely difficult.
I guess the easiest way of explaining that is if you can see that I have white hair and that's been caused just by fighting with people all the time.
I think it's been a combination of things which have been difficult, for example, actors have gotten so used to doing ADR and dubbing.
As one actor told me: "I can act or I can deliver the lines, you tell me which one you want" and most of the problem is, in most of Bollywood, the stars are not actors, so I remember telling some of the big stars: "Maybe I should have got you to go through the lines because you're such a terrible actor in hindsight".
What has been difficult has been also that now the younger directors are hipper to it, but the older directors didn't want to change things.

It was also about the discipline coming on to set, which was very difficult.
In India it was difficult to shoot from real locations and the fact that we did not have any soundstages, they were just a shell with terrible acoustics, but that's how it was, and yes, it certainly has been a battle.

Edgar
In how many years has this process been moving towards a new way of recording sound, what kind of time frame?

Nakul
I'd say that not all films are done with location sound even today, there will be some films which are dubbed, but with OTT platforms having come in and more and more being used, it's a necessity where you have to do location sound, so that's the good part of it.
But from 2000, which was when *Lagaan* was done, I'd say by about 2014 or 2015, we started realising that a lot more were being done. Even today they're not doing this role very well, they do leave a lot to be desired.

So despite of using all the best equipment, I think that's because there hasn't been a mentoring role, so like sound guys haven't taken on an assistant who has studied in London, but someone who has experience and then gone out and build it up or being a boom op first and then changing.

You are a musician and you have many years of experience in the studio. Has this background played an important role in your work as production sound mixer?

Yes, I think so.
The passion for sound comes, I guess, from really being a musician, a failed musician and a better engineer than a musician.
Definitely, that has helped.
It also is something which then keeps you ahead in terms of technology and reading out about new microphones and new toys, but most definitely, having come from a musical background and being in the studio, it has helped.

Edgar
So in terms of growth as well, in terms of having an ear for sound, has that been helpful?

Nakul
Yes, it has, because initially I started off by doing a lot of television work, commercials and jingles.
I counted them once and I stopped at about four thousand five hundred television commercials.
Those were the days of U-matic, Beta machines and cameras, and TV suddenly came in at a much cheaper rate and it was quicker turnaround; faster turnarounds and advertising boomed at that point in time.
Yes, I think that's where a lot of my attention to detail came from that because you had 30 seconds which we would cut in to 24 or 25 frames and so each frame was important. Unfortunately, we didn't have Pro Tools and things, we just had Fostex G16, 16 track analog recorders, so which meant your decision making was more immediate because you had just that many tracks to do things on, that would be voiceover, singing, music, sound effects for the commercials, so all in 14 tracks, because one track was timecode to locked U-matic tape too and one track was to trigger off machines with. Nightmare.

When you record sound for a movie, what is your main objective? What do you try to provide to the post-production team to ensure a scene is complete (from the point of view of the sound recorded on set)?

It's always got to be the best, cleanest possible dialogue.
I couldn't care about anything else.
I mean, effects and things can always be done later, and as you know, we put down mats so that we don't have footsteps. If there are any particular effects, which are there on location, either I record them during the shot or I will go back record them separately so that post-production can use them.
I guess a lot of this has been also since I do my own Post, of whatever I've done on location as well.
It was a very quick learning curve that dialogue is key and everything else comes second, but it's nice to also have room tone, it's important when you need to ADR lines or stuff like that.

Edgar
So we can say that your main objective is the dialogue, and if on set you understand that there's something that is very important to you, be it the location, the ambient sound, effects, then you will deal with that, like you said, before or after.

Nakul
Yes, I'll make my director aware of a particular sound that I would love to record and then we'll figure it out when to do it. Did you see *Lion* recently?
On *Lion* when we were on the location recce, I asked the villagers for something particular sound wise, and they said that there was a tree where the bugs would come.
I had a little two track recorder with me and I said, "I'm going to go, I'm going to stay back and record it", and I did.
Then when we were shooting there one day, they were doing some drone shots and I said, "OK, well, let me go and try and see if I can record that even better with my whole rig".
My director asks me "Where are you off to?" And I said "Oh, there's this tree which I'm going to…" And he said "Oh, cool". And then I think somehow he came back and he called me when he was doing Post and he said: "I need something", and I said, "I've got these bugs, why don't you check them out?" and send it to him and bang, that's what that's in the movie whenever he's having the flashbacks and going back to that

place, that's the sound which takes you, that leads to that.

You always listening for things, in any case, you're always listening to things that you want to take off.

I remember in the film called *Dil Chahta Hai* the assistant directors went insane because we were shooting in Bombay and in those days, there was a lot of construction and I went and said: "Stop that work".

And when the movie came out, I had replaced in one particular scene the same sound of that construction work, and they said: "What you got, you made us take it off what you might as well have, just let us leave it on", and I said: "No, but the dialogue was clean, which is why I could add that later on. With the dialogue it wouldn't have been the same".

Edgar

That is because you have the experience to know when something can work and cannot work.

Nakul

Well, that was my third, so I think it was pretty evident, I had to be a quick learner.

Edgar

When we you mentioned dialogue, do you mean also the off-camera dialogue, is that also important for you? It needs to be very clean?

Nakul

I always use two booms; one is always on the primary actor and the second boom is always on for cues and secondary actors and things.

For the most part, unless something is written in the script, I don't like off camera things being said by people, because if they're not in the frame I don't want to hear them. I will record that separately as a wild track later on in the same ambience, in the same environment, with the same microphones.

But also more importantly, I think, is when people are talking and a lot of people say just say something and it makes no sense because it's not relevant to it, if it's got any relevance to the scene, then yes, I will record it then.

I mean, I count that as something which is needed but if it's just garbage, like "Hey, where are you going?" You know, any normal rubbish then I can group that or I can just get that quickly by myself separately, later on.

133

Nakul Kamte on set of *Harud* directed by Aamir Bashir - 2010

When you read a script, what is your approach toward the sound? Which types of scenes worry you the most? What is your next step after reading the script?

I break it down, into the night, interior exterior, all of that, I think what really, really comes to me is scenes, which I know, a soft, quiet, emotional, because those are the most difficult to do ADR.

On the scene where if someone's crying or breaking down or whatever, no matter what, I mean, even if a bomb falls, I will never shout "cut" on that scene.

I will just let it roll and let the director take that, because I know the actors, how difficult it is for the actors to work themselves into that.

If it's a more casual scene, I would ask for a cut, but if there is a noise, which is, you know, if it's like if there's an accident going on or something or whatever; this is India, there's always something going on.

Edgar
So first, you said that you read the script and then you break down the scenes for you to understand which ones are day or night, inside or outside, so you make notes, obviously, and what's the next step?

Nakul
Well, if there's a thing where they'll be saying this is in a crowded market or whatever, I highlight that, so I know that I need to record some market ambience for that scene.

During the recce, I'm thinking: "OK, if we're there, if it's a set or if it's a real thing", one of the things we try to really, really pay a lot of attention to is where the generators are going to be put.

Like they say, "silent generator", I still have to hear a silent generator, I've never heard of a silent generator because you can hear them, which is why they call them silent generators...

Edgar
They're always too close no matter what.

Nakul
Yeah, so I actually get them to carry extra cable with them and sometimes you have to fight for that.

But I think one of the keys to me, what I found is that you've got to be on good terms with your DoP because if he understands where you're coming from, in terms of passion for the film, he will do things to help you.

I mean, I remember in *Lagaan*, there was one scene where the DoP was setting the light, we stopped for dinner and we were supposed to shoot immediately after dinner and the wind picked up and he said "No, I need one hour more".

And, well, he'd already got it ready, so I walked the set, and he was changing all the gels and I said: "What are you doing?"

He said: "The wind is coming down and is going to make noise on the gels and you'll complain".

So, I said: "You know what? The dialogue is after you've done this track and it looks beautiful with the way you lit it, but if you're changing the gels, I can tell it's not looking the same, it's not the same thing".

He said: "So you're OK with that?"

So I said: "Yes".

He says: "No, no".

I said, "Don't worry, it's one line and I let it go, just keep you, and you need to make it look like this, because that's what the flavour of the scene is. I will get my one line, don't worry about it, I'll take a wild track".

And after that, I mean, he was mine.

He realised that I was willing to sacrifice things for him and he was always on the lookout to help me doing it.

There has to be a bond between the sound guy and the DoP. Unfortunately, most DoPs think they're God.

Anyway, we won't go there.

Edgar

And so the scenes that you would be most worried about, what kind of scenes are those?

Nakul

Quiet scenes, emotional scenes, scenes where ADR is almost impossible. You'd never get the same feel.

Normally where people break down and start crying because then to get that, that's really difficult to ADR properly.

Having said that, if I feel that using ADR can improve a scene, I'm great with doing that as well. It's actually whatever works for the film, it's all about telling the story right with sound.

How many people make up your standard sound team? Which characteristics should a good boom operator have? Do you personally place the wireless microphones on the actors or is it responsibility of your boom op/sound assistant?

I would love to have a larger team, though I'm not really sure about that.

I would like to have one student on my team and in the last two years that's what I was doing.

A standard team because I don' t know what it's like in Europe, but over here in India, it's always dinero, more budget, so my team would be me and two boom ops.

Boom operator essentials: change their socks every day, brush their teeth, are presentable, smell nice, have a polite disposition, don't get into arguments with actors or directors or DoP's about booming field.

What I do like about the boom operators is if they are always on the edge and the DoPs are a little scared of them: I'm about 1.92 cm, my first boom op is around 1.92 cm and my second boom is 1.88 cm, so we threatened the camera department.

I sometimes wire people myself.

I'm happy with the way womens' things are going especially on certain films and certain actresses, I've had to say that they know what they're doing…I'm not going to argue with Maggie Smith or Judi Dench, I just put it in a Ziploc bag, make sure that it's clean, that there's new tape, that there's everything in there, so she just puts it on and if anything is asked I say: "Yes, ma'am is good on, it'll be little adjustments here and there".

But she's been doing this for 40 years, so she knows what she's doing.

With Indian actors, they are a nightmare.

Some of them will purposely leave the radio mike in the Vanity Van and come to set without it and say, "Oh, sorry, I forgot", not with me, but with my assistants and things, I think, again, because being my size and being a little older, having a reputation of calling a "spade a spade" and really telling them, well, in any case, it does. I have said on occasion, very rudely, I must say: "You know what, you are deliberately so fucking pathetic that I wouldn't have made a difference if I had a mic down your throat".

Nakul Kamte on set of *Hotel Mumbai* directed by Anthony Maras - 2017

137

I remembered one show where we had a great actor called Naseeruddin Shah and there was this actress called Deepika Padukone for a film called *Finding Fanny*.

We were shooting in Goa at night and so she's supposed to be whispering to him, but the crickets are so loud that I couldn't hear her, and I said: "You've got to say it out loud".

She said: "Put a mike" and I said: "I've got a bloody mike on you and it's all over it, so what do you want me to do?".

I can't do this and whatever, and the director said: "Calm down, break for dinner" and I just said: "Dude, I'm not doing this to go home and listen to this, this is for your film, kindly explain that to her".

We finished dinner then I got in and Naseeruddin realised what is my problem, what I was going through and Naseeruddin is one of the most, one of the best actors we've had in this country.

He's very senior and he earns a lot of respect.

He just looked at me and I just I was still shaking my head and he said: "Deepika, I can't hear you, so I don't know how to give cues," and she said: "You men always stick together".

I said: "No, it's not you men, it's actors and sound people stick together because he doesn't want to dub it, because later on you would ask me: "Why am I dubbing this scene and why is he not dubbing this scene?", and it becomes hell so understand that we are doing our jobs also.

Sometimes you have to be brutal when you explain it to them, it's a bit of massage, it's a bit of thing, it's a bit of sometimes being angry with them.

Sometimes I've thrown off my headphones and said: "Screw it, remember your dialogues, we'll see you do this in a dubbing theatre" and walked away.

Of course, I've kept my assistant who's recorded it after I've done this and gone for a cigarette or whatever.

Edgar

Do you think that in order to become a good production sound mixer you have to be a boom operator first?

Nakul

If you don't know that if there's a glass table and they are booming, how if you angle it wrong then it's going to bounce off the glass and maybe you should know what you're doing with it.

But then again, I think everyone has done booming in the early days, I mean, if there were certain shots which were on a Steadicam, I would have to boom myself because there was no way you could follow anyone. I mean, how do you do it with cable in those days? There was no wireless.

But yes, I think it's necessary to know a little bit about booming for sure, because only then do you know how it should sound.

Based on your experience, do you believe there is an "Indian", "American", "European" style/approach to this profession?

Well, it's noisier here than it is in America or in Europe.
I find it a lot easier to record in those countries than here.
The other thing which happens in northern India is you attract massive crowds who will be "bebebebebebebe" throughout, it's a nightmare.
But I think for the most part, it's pretty much the same, everyone uses the same equipment, the same microphones.
I think now that I'm working at Netflix, one of the things I re-

alise is between the good production mixers and the ones who I don't consider production mixers and that I'm realising in whether it's not so much in America but even there, Europe a little bit, is gain staging of what they're know, how they're using gain staging on the mike preamps because if it sounds shitty and if you've got a really bad quality source, then there's nothing you can do with it.

It's also a question of budget, I mean, over here, people, if they're not going to give you budget where you can have Lectros or Wisycoms or Zaxcoms or whatever, and if all the budget you've got is for Sennheiser G3 o G4, your radios aren't going to sound good.

Budget is a big problem over here.

Edgar

But overall, based on your experience, the way this profession is done, the way this profession is carried out, is more or less the same in different countries or do you see also a difference in the way you work in India or in other countries?

Nakul

I think there is a little different way in which we work.

We have a thing what they call "sync sound security", which is basically goons, tough guys going telling people to shut up.

And I have no issues with them going and doing what they need to do to keep it quiet, get the set quiet for me.

I wouldn't think or daydream of doing that in Europe or America because we'd get sued.

The second main difference is about the background.

Over here, they're extremely badly behaved, in the West they know what they're doing. They are happy to have the job. Over here it's like they're struggling and they're so frustrated that they will always be talking.

A couple of them now, they know me and they know how to behave on set.

Edgar

It must be extremely stressful to be a production sound mixer in India.

Nakul

Definitely.

140

Nakul Kamte on set of *Lion* directed by Garth Davis - 2015

The relationship with other departments is crucial. If everyone worked not only in the interest of thier own department but also helping each other, everything would be easier and more fun.
Have you ever found yourself struggling with other departments? Any interesting anecdotes you would like to share?

You have to be the best of friends. As a production mixer, if there's one person you make friends with on set it's the director of photography because if he hears it from your perspective, he will start listening to the silence of what he's seen.
It becomes magic and it makes your work really easy, or comparatively easier.

I got to love grips and gaffers. I was very lucky on my first film, I had a gaffer who's a legend.
I'll never forget him telling me, he said there's always one position, if you can't find it, I will tell you where to boom from. No matter what it is, even if it's a 360° lighting, I will find you one position to put your boom.
Grips are phenomenal because they always help you to find the right position.
Hair, makeup, wardrobe is very important to me also.
Hair for those to get into little clips and see if I can do a hair-mount.
I was lucky that I did some plays, so I was used to working with people who would do hairmounts because in theather they do a lot of hairmounting, so if you've got a good head person, that's one huge advantage because it's straight down, absolutely clean, no rustle, it makes life really good.
Wardrobe, similar stories.
During pre-production meetings, I talked on and on and on about what fabric to use and not to use silk and not to use other fabric, which is noisy, but do they listen? No, of course not.
They're never going to listen.
You're just stupid sound facts, why should we listen to you?
You only have 20 years of experience. And this is my first film…

Our relationship with the actors and with the director are crucial to perform our job successfully.

Both can help us get good sound and we often need to make demands of them.
How do you relate to actors and directors?
What should the right profile of a production sound mixer be?

I think you have to make friends with both of them and invariably one, either the director or the actor is an asshole. Very rarely will they both be nice people.

So, it's all about then figuring that out really quickly and using the good cop, bad cop thing.

For example, I recall when I was working with a terrible director who didn't give a shit about sound, until he suddenly realised and he said: "Don't talk to the actors, don't even look at them, just roll, don't cut".

And I said: "Wait a second, sorry, sorry, sorry, I can't do this. This is not how it works. This is not how you fucking make a film".

This is like, I teach you.

He was a professor at Columbia University where he taught filmmaking.

I said: "I'm sorry, I mean, you cannot be doing this because this is not how you make films". "This is the way I want to make movies".

And I said: "Well, fuck off, get someone else then, because this doesn't work. No one knows what the fuck we're doing. Everyone's asked me to stop, and we've shot for three days and we're already two days behind".

Production was: "Oh, man, we were waiting for you so much". And then, of course, I apologised and I made friends with him, and then he realised what I was wanting to do.

Then he also realised that a lot of it was in Hindi, you know, and I warned him, I said: "Look, you are doing this where you've got Indian actors speaking in English and in Hindi, now the Hindi, I can tell you where you have to listen to me because I'm yours ears, but the English what I'm warning you about, which you're not realising is that this speaking really fast because Indians tend to speak English really quickly", and then sure enough, he says: "I want to ADR this line and this is going to be out of sync, and Netflix said, no, you can't ADR…".

So, it's a mess.

But like which is why it has to be a combination of working with the director and the actor.

Edgar

And so there are directors, indeed, that are more interested in sound than pay more attention to it than some that are not, and so when you work with some directors who are interested in sound and some who are not interested in sound for you, do you change your approach working with, for example, a director that you know is interested in sound compared to one that you realise he's not he or she?

Nakul

Yes, there is a change in approach with a director who's interested in sound, he has his ears open, he's listening and he'll ask me to record this or other things.

With a director who's not interested in sound, I've got to think of all of what he should have been thinking about and record all those things.

So that when he gets to Post, he goes back embarrassed and says: "Thank you, without you I'd be fucked".

If you're passionate about sound, you do that.

Tell us about your equipment and how you have organised your sound cart.
Is there anything in your sound kit that you could never do without?
Nowadays does it still make sense work with the cable (for boom)? Yes? No? Why?

Good question.

Well, I use a Grace Design mic preamp, which is why my boom sounds so beautiful.

But more and more, it's wireless, I think everything is wireless because now there's just so much run and gun and 360s and stuff, but I still don't think it sounds as good as my Grace design.

Standard kit used to be a Deva 5.8, started off with two DAT HHB and Fostex PD6, went to Deva 2, Deva 4, 5.8, Nomad, I've used the Cantar, Sound Devices, Lectrosonics, Wisycom, Zaxcom, the whole thing.

I like the sound of Schoeps for indoor, but it depends on if I'm shooting through the monsoons, then I don't use the Schoeps then I'd rather use Sennheiser, simply because the moment there's moisture on the Schoeps…

Edgar

So, you change your equipment a lot

Nakul

I carry about 20 microphones with me on set.
And everyone is like: "They don't pay for all of them". So, I
said: "No, they don't, but it's my sound and that's it at the end
of the day, that's my name on it".
So I try and get the best rate I can.
There's no one mic to do everything.

Edgar

So your current set up. What is it right now?

Nakul

No, now I'm doing very little recording because I've joined
Netflix, I'm a sound specialist with Netflix.
There are now five of us globally.
When I joined, I was the third person: two in LA, me in India,
one in Japan and one in Europe.
What I have left is a Zaxcom Nomad, 12 channels of Lectro,
so that's two wireless boom and 8 or 10 radios or three wireless booms if I want.
Couple of channels of Zaxcom because I love them, because
if they go out of range on the boom it keeps recording on to
a card and so you don't lose any audio.
It's also great for car rigs and stuff, even if they go out of
range, it's still being recorded and later on you have all the
audio.
I love my Cinela mounts. I think it makes a great shield, so I
like that.
What else? I've got so many different little odds and ends,
which I've picked up through my travels through, everything
from Rycote stickies to Viviana, to BubbleBee, to all of them,
I guess just having a variety of different things for different
applications and needs.

Edgar

As a lavalier?

Nakul

DPA4060, Countrymen B6 sometimes if I need a really
small profile thing, Sanken Cos 11. Yeah, mainly those.

144

Edgar
As you stated, you would prefer working with cable, but nowa-days it is getting more and more difficult. That was your assessment.

Nakul
It's also getting closer.
But there's a certain… it's like listening to a record and listening to FLAC; the record still sounds like a record and there's a certain warmth to it, even if you've done FLAC, which is high resolution and all, there is a difference.
It might only be to me, but it is to me and I'm the sound guy who's recording it, so I want it to be the best that I can possibly.
One of the things which I do is with certain ambiences and certain effects, or even with voices, if I know that that's going to be a scream, I might actually record that at 192khz or 96Khz just so that if I pitch it up and down the less artifacts, especially for gun shots and cars; I love doing that also.

In your opinion, how much of an "artistic" side and how much of a "technical" side are there in the work of a sound engineer on set?

I think it's both.
I've never stopped to really think about it because it's just a case of doing it, because if you stop to think about it, then there's something wrong.
It's got to be organic.
You've got to just let it flow.

Edgar
So something that has to be automatic, switching from one side to the other, there's a golden medium in between.

Nakul
Yeah, I think the experience teaches you that and also the thing about being a production mixer or whatever, I mean, you're part of the crew and you're hired because you deal with issues and problems, so they don't seem like issues and problems, no?

When we talk about this job, we seldom address the importance of protecting our health, our body and of working safely.

You did it in one of your interviews and it is truly appreciated. Any advice you would like to share in this regard?

I put down my faders after 12 hours.
I would now say if I was starting off again, I'd say I tried to do it at eight or 10 hours.
You only have your health, you only have it once in your life.
I think also looking after your spine and your back, if you have it on your chest or something to think about and consider. I know it's something new, like the the Zaxcom Nova it's fantastic because it's so small and light and you've got digital things built into that and it sounds great, but I think health is a very major issue and I think safety is something which now you're starting to see.
I can't speak enough about looking after your health and don't let production bully you into that.

For a young sound guy or girl who wants to pursue this career, it can be difficult to take the first steps, gain experience and at the same time ask for proper remuneration, however working for low fees can damage the entire profession. It is difficult to strike a balance. What advice can you give to young people at the beginning of their careers?

Unless you're fucking crazy, don't fucking do sound.
Unless you love it with the passion, don't even fucking think about it, do something else, get smart before you go down that road.
No, but all kidding aside, now you've got to have passion about it, and I think if you have the passion and you speak to other sound guys who are a little older who have been around, they will pick up that you've got passion.
That's what I do with a lot of kids, the moment the ones who chase me, I have them everywhere I can, I mean, I show them how to use stuff, chat with them, hang with them, tell them stories which they might learn things from, I'm happy to say that about eight of my assistants who are now very successful sound recordists.
So that feels really good.

Edgar

And *as far as fees and pay is concerned, what advice would you give to young people when they start off? What kind of remuneration should they ask for?*

Nakul

Do not do it for less. Ask around.

You're not going to get a big film to do without any experience, so you're not going to be getting that those kind of jobs. Don't do anything for free.

Try and get the most you can.

Always try and get the best equipment you can.

Do not believe "This is going to look great on your resume" it's not, that's bullshit.

Your own industry is going to hate you if you do that, because if you're undercutting, then you're also shooting yourself in the foot.

I wish there were unions like the Americans had, which you'll get paid really well with, but at the same time, there are also non-union films, which a lot of them are.

You've got to be true to yourself at the end of the day, also realise that a lot of people will say, a lot of film school students say: "I want to do this and it's for the magic of cinema". That's bullshit.

The magic of cinema doesn't fucking put food on the table and feed your family.

That's the bottom line.

That is the magic of cinema, which I want to see.

That's when you'll start charging and thinking about yourself a little more serious role.

Patrick Becker
Switzerland

Patrick Becker

Patrick Becker was born in Brussels (Belgium) in 1966 and now lives in Basel (Switzerland). He attended primary schools in both Italian and German-speaking Switzerland. In 1986, he enrolled at the Academy of Fine Arts in Basel, which he left after 5 years with a diploma in audio-visual design. Since 1992, he has worked as a sound engineer for documentaries and feature films. In 2002, he was a lecturer at the Lucerne University of Applied Sciences and Arts, Video Department, and in 2005 founded the company "NurTon GmbH", a studio for audio post-production. Since then, he alternated between jobs as a set sound engineer and film sound post-production.

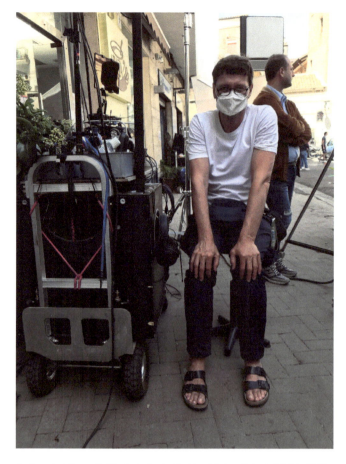

On the set of *Presqu'ile* (*Au Sud*), directed by Lionel Beier - Catania (Sicily) - 2021

What does it mean for you to be a production sound mixer? What are the pros and cons of the life of a production sound mixer?

Being a production sound mixer means working on a project together with others.

It's important to be able to fully concentrate on the common production process, to be determined and precise to implement the goals that you have set yourself with respect for the work of others. Film shooting is a common process, a shared development. The goal is a more or less successful collection of recordings converted into images and sounds, which are then reassembled later.

Being a production sound mixer means that you not only have to have an idea how to implement the scenes but also the content that lies between the lines from a script. It means that you have to find good, simple solutions on how to get the most out of the various shooting situations. I see myself as part of a larger team, the entire audio team working on a film project, and when I record beautiful sounds, unexpected noises, touching music or simply hear clear voices, it always makes me happy.

Of course, as with any profession, there are also negative sides, a bad script, unfriendly actors or impossible locations where a good location sound is not possible.

Stress has perhaps become one of my greatest enemies over the years.

It is unnecessary and always arises when you want more, in my opinion.

Annoyance and dissatisfaction also comes from a dubious preparation by myself as a sound engineer or by the production, this is then often the reason for ambiguity and chaos with many unanswered questions for which you don't have time at the moment of shooting. Other reasons range from actors who don't like being wired are also unpleasant, long night shoots, small crowded locations, too much wind, too much cold or heat or too much dust.. and again, very long travel times or bad food. Also a hateful mood on the set or arguments.

But who wants that? That's why everyone usually make an effort to ensure that this doesn't happen.

How much attention is the Swiss film industry paying to location sound recording today and how was it in the past? Has there been an evolution from this point of view since you started working as a production sound mixer/boom operator?

Yes, since I started working as a location sound mixer, there has been a general increase in awareness of the consumption of audio and visual media. And the various funding institutions have increased their funds. The amount of movies watched has increased sharply. Making films has become a lucrative and respected business, especially when the work reaches a large audience.

But how can independent filmmaking exist alongside the commercial interests of productions optimized for audience numbers? How can protest films or uncomfortable films or incomprehensible, extravagant or even nonsensical productions be as well funded even if they probably won't bring in any quantitative success but are still quality productions? What is creating quality in a film anyway? This is decisively decided at the time of the funding.

Series as a format have taken audio visual stories into a new dimension. But how did the audience and their content needs develop? I think it's everyone's responsibility to educate themselves culturally. Film is such a cultural product and it is our responsibility to determine what we want to see and what we deal with; even if that depends of course on the offer and its diversity, it also applies to the quantity, how much we would like to consume, what do we need and what is good for us. Regarding the production landscape, I believe in an educational mandate that the various funding institutions should take on, so that education and development can generally take place through the discussion of socially relevant topics.

In terms of my job as a sound engineer, that has actually developed positively.

There is more money and you want to implement a lot more in a shorter time, whereby the more advanced technical means really help you.

How much stress arises on a production is usually chosen by yourself.

Swiss producers and directors have always been attentive to the interests of the sound department.

As the technical possibilities have evolved, so have the demands and complexity of the equipment used on set to record sound. Often it is necessary to use two booms, eight radio microphones, a stereo pair, a second recorder to be placed in a car for the same scene, as well as perhaps dealing with playback and recording. All this in addition to the radio headsets to be distributed to the other departments, and without forgetting the management of all the timecode systems that allow audio and video equipment to be synchronised with each other.

Making all this come together is a lot of work and takes time. Times have changed, nowadays you can do and would like to do a lot more, but what you have to keep in mind is that this must also mean more personnel, so it makes total sense to have three people in the sound department, and all the different recording situations should be planned precisely. Perhaps not all production companies today understand why they should make this extra effort and finance it, but it is worth fighting for and you have to communicate and weigh up exactly the advantages and disadvantages.

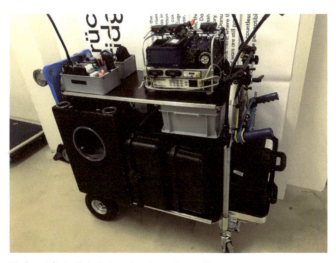

My Sound Cart with batteries, microphones, transmitters, accessories and a Boom Box

On the set of *Semret*, directed by Caterina Mona - Zurich - 2020

When you read a script, what is your approach toward the sound? Which types of scenes worry you the most? What is your next step after reading the script?

When I read a script, I first try to recognize a core mood in the text to be able to better imagine the different implementation options for the sound work in general throughout the entire film. I'm particularly interested in the possibilities and difficulties, which are already announced in the script, for the dialogue recording in the various scenes and recording situations.

I note how many actors are involved in a scene and which sounds I could still record from the scene on location. I write down special recording situations such as car trips, concerts or phone calls to later discuss with the director or the production how we can best record them.

Whether a scene is difficult or easy is often determined by small details such as the time of day or the noise of birds etc. or whether children or animals have a role on the set, where the recordings take place and whether there are loud noises that appear as motifs in the script or even which do not occur but are present at the location and would be disturbing. These elements, which are important for sound recording, need to be identified and planned as precisely as possible so that you can take as much as possible from the original sound from the set. Even if you want to recreate everything in post-production, the original shooting location is a very important source for the timbre and quality of the original direct sound.

The greatest difficulties for me arise when several elements come together that make recording sound difficult. For example, combinations such as quietly spoken dialogues recorded at noisy locations or unpredictable spontaneous staging decisions for which there is then no more time to take them with the appropriate means, because, for example, the sun is going down and soon there will be no more light, and so on.

Actually, the more I can anticipate for a recording situation, the better I can plan and implement its sound recording.

When you record sound for a movie, what is your main objective? What do you try to provide to the post-production team to ensure a scene is complete (from the point of view of the sound recorded on set)?

I would like to record a clear sound that also fits the perspective of the picture, or at least offers the possibility to be adapted later to a desired perspective. The lavaliers or plant mics are the miracle cures. It is certainly the main task of the sound engineer to record the dialogue in all its nuances and subtleties as the actors modulate it. However, it's just as important to record further sounds and dialogues that might take place in the background within a scene or that specifically characterise the location or a mood and only allow it to be immersed in a desired atmosphere. These can be individual conversations that you don't necessarily have to understand, individual screams or laughter or the talking of a whole group of people called "walla". Then again, it can also be neighbours who walk around on the upper floor and do ordinary everyday things.

Quality also means separating the foreground sound from the background as far as possible, because only if each sound source is recorded as separately as possible from the others can you then edit them individually in post-processing and reassemble and arrange them again.

Therefore, it's important to record all additional sounds for a scene separately.

These can be departures, passing or arriving cars, or doors and noises from other actions that overlap scenically with the dialogue. These can also be standing noises like wind or refrigerators, traffic or other noises that are present in the room.

In order to be able to find your way around all these recordings in the subsequent post-production, it's important that these recordings are clearly labelled: notes should specify whether they are additional sound recordings like wild tracks or something else, or whether these recordings were recorded parallel to the picture.

I usually write down the location and the subject if it's something other than dialogue.

158

What features should the sound you are recording have to make you happy and satisfied, and vice versa, what could lead you to be disappointeed/dissatisfied with the sound you are recording?

Actually, everything that pleases is also good and beautiful or satisfying to record it.

Of course, particularly beautiful voices or interesting sounds are also correspondingly a pleasure, or the intense play of an actor, a song or simply an interesting noise with many nuances or a great mix of sounds. Music is always a pleasure to record. When recording sound on set, I am simply happy if I can record the individual sound sources sufficiently separated, the foreground from the background to a sufficient extent.

A particularly interesting aspect on film sets is working with the actors who build up a character and put emotions in their lines and figure out through technique how can I best record these emotions in different situations.

But it is also the quality of the script, how the images and sounds interact together, which is then reflected in the sound recording and makes it interesting.

What annoys me is when it's difficult or almost impossible to place radio microphones without them being noisy. For example, because of the clothes themselves which rustle a lot, or a combination of clothes creates friction or are simply too noisy. I also find it annoying when actors speak so quietly that they almost whisper and you can only hear them from a distance of 20 cm. It's also not pleasant when car or rain noise covers everything and makes recording difficult.

Do you personally place the wireless microphones on the actors or is it responsibility of your boom op/sound assistant? How many people make up your standard sound team and with which tasks?

I usually wire the actors.

I've just seldom seen assistants doing it the way I'd like them to. Apart from that, it is not simply placing a lavalier, because you also have to check how it sounds and in the course of the shooting you should always check and adjust the correct placement or remove the rustle if there is any.

The assistant (Perchman or Boom Operator) is often busy with the dialogue or with light problems before recording a shot in order to memorize the best way for the planned shot.

If there are a lot of actors, however, I'm happy to get help and if the assistant can do the wiring, I am happy as well.

I also like to record dialogues with a second boom in order to pick up also small noises from actors. Working in a three-some is actually my ideal situation. Some scenes can certainly be recorded well with a production sound mixer and one boom operator only, in other scenes it is often useful and enriching to record them with two boom operators.

If there are three people working together, it's also easier to record additional sounds, so that during a simple scene a second assistant, for example, moves away from the set and goes to record atmos or wild line with actors or cars or other objects.

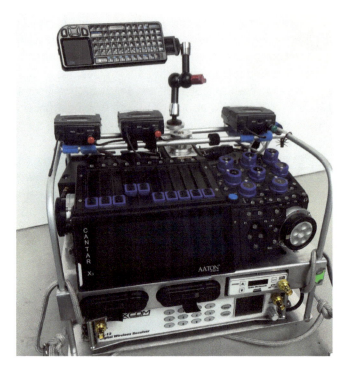

My portable compact Rack for small locations

Can you give us an overview of your professional tools/ equipment? How have you organised your sound cart?

Both my main and second recorder is a 24 track Aaton Cantar X3 with Cantarem 2.

I have an RX12 from Zaxcom with IFB200 as radio receiver. I have eight medium-sized radio transmitters TRXLA3.5 and four smaller radio transmitters ZMT4 as well as three plug-on transmitters TRX743 for wireless booms and plant microphones.

For microphones I mainly use Schoeps Super CMIT on the boom and CMC6 or CCM for stereo setups and plant mics. I use DPA 4063 with a Microdot adapter as a lavalier, Sennheiser EW 100 transmitters and receivers for monitoring purposes.

To get a picture I have a Teradek Cube 655 with Link that sends the picture to my iPhone or iPad via WiFi from the camera or from a video village.

All recorders for sound and picture have an external time-code generator from Ambient Recording and are connected together on a WiFi network.

The recorder and wireless receiver are built together in a portable rack so that it is as small and compact as possible to transport it in a vehicle or carry it around, which isn't often as it's a relatively heavy 15 kilos.

Usually, I place the recorder on a small magliner on which I then mount the Shark Finn antennas from Zaxcom. Otherwise, I still have with me a lot of windshields, furs, tripods, boom poles (in different lengths), carpets, cables, adapters, batteries, headphones and work clothes for the rain and cold, a parasol or umbrella, a large loudspeaker for playback, small loudspeakers and much more.

Then I need an iPhone, an iPad and a computer as well as hard disks for the backup and headphones! Beyerdynamic DT1770; I love them...

Nowadays, in 2021, does it still make sense work with the cable? Yes? No? Why? Can you give me some examples?

Yes, absolutely, if I compare the same audio signal from a microphone, one via a wireless system and one transmitted via cable, I notice that the dynamic range is significantly greater with cable and at the same time I can work better with the preamp and not just with the volume at the end.

In many shoots, the boom operators are more or less stationary in a specific place and the boom itself is also lighter to hold if a transmitter does not have to be attached to its end as well.

In many other situations, however, the cable is a hindrance, and it's absolutely necessary to be able to use a radio transmitter when movements are complicated or distances have to be covered.

What is perhaps more practical is that digital signals are transmitted much more interference-free through a cable, so it does not matter if they are placed over power cables to which a dimmer is connected.

What is also relevant is that with a microphone connected to the recorder with a cable, I have a larger gain range and the better noise floor in order to be able to record the best possible audio signal adapted to the sound source in terms of level, but maybe that has just changed with the new Shure system.

The relationship with other departments is crucial. If everyone worked not only in the interest of their own department but also by helping each other, everything would be easier and more fun. Have you ever found yourself struggling with other departments? Any interesting anecdotes you would like to share?

Well, they might not be anecdotes when the camera decides to use a loud smoke machine or the costume department has chosen very nice but extremely noisy clothes for actors, or when the lighting lamps are placed so close to the actors that you can hear them, or if it is decided during a car scene that the window should now be wound down to avoid reflections although, according to the scene, it should be closed.

However, no one does it on purpose so, once again: preparation, information and communication is everything. A location recce can often be very informative in terms of how the shooting day will be in terms of sound.

Sometimes there are situations where you really have to help each other because everyone has already worked so many hours or because the circumstances are so adverse with rain, cold or snow.

Camera Car setup with too short Antenna cables!

You have a post-production studio and often in your films you take care of all the phases, from sound recording on set to sound editing to the final mix.
Is this your favourite workflow (to do the whole sound process by yourself)? What do you think are the pros and cons of this type of working?

163

I really like relaxed shooting situations, I really like feature films but also documentaries. The type of work on the set is very different from post-production because you are together with many people and in many different places, you will have many encounters and face a lot of unpredictable situations. In post-production you are often alone with the computer working your way forward layer by layer. This layer-by-layer work can be very satisfying when you clean and filter a dialog or get a good dialog separation from effects.

When you create effects or add ambience it is very nice to hear how the complexity of the soundtrack increases as the various elements are put together and grows into a deep emotional or realistic element in the film. But it's also a lot of graft if you work on an entire film alone, it might be too much work because you can't keep a clear head, and distance yourself over such a long time.

I also like to work with other people in post-production so sound editing, sound design, foleys, ADR, and mixing are often separate work steps that are often processed by different sound engineers. What I also see as a very big advantage, even on the set, is when sound engineers have experience in all these areas; not only because it's fun just to talk about all possible working methods and techniques all day long but to discuss with each other and make the best choices.

I think that being a sound engineer for a documentary is very demanding and that the sound behind documentaries is always a little underestimated or little appreciated. A documentary on a sound level can bring considerable challenges and, in the end, a lot of satisfaction.
What are your thoughts on this? How does your approach to work change when you are dealing with sound for documentary compared to sound for feature?

Yes, collecting sounds and recording conversations for documentaries is a completely different way of working than recording dialogues on a feature film set. In documentary films, you often don't know who'll speak or what exactly the actions will be, so you have to prepare yourself, be ready and weigh up the possibilities with forward planning.

Most of the time, you are often part of a small team and everyone influences both the content and the individual protagonists only with their presence, this is a highly sensitive situation.

I find it particularly important to decide on the set which sounds can be added in post-production and which are not so easy to find. Typical location-specific or mood-specific noises and sounds often make up the aspect of originality in documentary films, and so is more important to include typical ambience backgrounds, along with anything else that might be of interest and help with flexibility in post-production.

At the same time, however, I also think that documentaries are designed and created in the same way as feature film, which are largely created in post-production with the possibilities and restrictions but also with the freedom that post-production allows.

Perhaps in the documentary films post-production one often shows too little courage for implementations that are a little more daring than the simple naturalistic depiction.

I am very interested in artistic expression, experimentation and design without losing sight of the subject and content of a documentary film.

On set it is incredibly nice and enriching to deal with a topic and to get to know people and also to involve yourself a little.

On set we might give our very best to record any small nuance of sound, but then movies are mostly watched on a computer or on TV. In your opinion, how attentive is the average viewer to the sound, or rather, how much do they really understand good sound quality?

It's true that we are relatively subconscious of listening, most of us. But still, hearing is a main sense of orientation in everyone's daily lives so I believe that we are all aurally very experienced beings, even if this takes place in part unconsciously. A good quality recording or a good sound in a movie, or a soundtrack that you like to listen to because it impresses you so much or it provides you with so much information aesthetically or through language, I think everyone perceives this.

I think we're all auditive people and maybe that's why there isn't anyone who doesn't like music. Perhaps listeners do not always notice exactly how something was made or what the exact elements are that make up a soundtrack, but they notice whether they are impressed and touched or not. I think it is our job to adapt the content as optimally as possible to the media that is used to watch or listen to something. But yes, films are best watched and listened to in the cinema theatre.

On the set of *Semret*, directed by Caterina Mona - Zurich - 2020

166

Epilogue

Here I am, back at home after having "travelled around the world".

I must say that these interviews have given me a lot of food for thought. They have stimulated me to try to do better and better in my profession, to not be satisfied and to continue along my journey.

I hope that for the reader, too, this book will have been a good travelling companion, interesting and useful from many points of view.

Happy shooting to you all, the best is yet to come!

168

Edgar Iacolenna
Photo by Giorgio Marturana

About
the author

Born in Rome in 1986, Edgar Iacolenna began his career
path with humanistic and musical studies.
He has more than ten years of experience in the film and
audiovisual industry as production sound mixer and boom
operator, collaborating with several film and television productions in Switzerland and Italy. Edgar has made several
recordings of solo concerts, chamber music, choral and sym-
phonic music, solo and ensemble of jazz music, as well as
theatre shows, editing the audio post production as well.
Since 2014, Edgar Iacolenna has been "sound" instructor in
the course of Audio-Visual Reportage led by Daniele Segre
at the *Centro Sperimentale di Cinematografia* (National Film
School) in L'Aquila (Abruzzo).
He has been living in Lugano (Switzerland) since 2015.

170

Acknowledgements

I really want to thank all the people who have helped me to realise this project of mine; first of all, my family where I am a son (Aurelio, Luciana and Giacomo) and my family where I am a father (Chiara, Alice and Duccio).
Thank you for your love, support and advice.
In addition to them, there are also many other people I would like to thank from the bottom of my heart: Michael Hoffman, Chiara Salce, Federica Martelli, Christopher Farley, Vera Bianda, Mena Torre, Lan Yao, Marco Monti, Gaelle Maurelli, Paola Sini, Julie Mucchiut, Oleg Magni, Michele Pennetta, Suzanne Tanner-Mosimann, Stefano Mosimann, Simon Manzoni, Filippo Toso, Alberto Meroni, Mariangela Marletta, Giacomo Jaeggli, and all the sponsors who believed in this project.

171

Advertising

Lockit Timecode
Sync, Workflow, Logging

www.ambient.de

experience
Quality

QuickPole
Carbon Fibre Boompoles

www.ambient.de

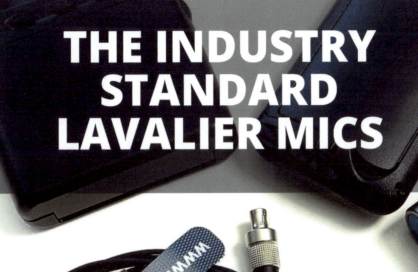

THE INDUSTRY STANDARD LAVALIER MICS

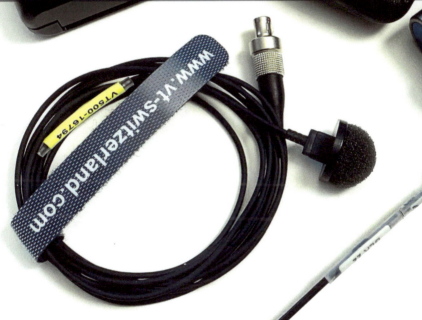

VT500/506 SERIES

Swiss precision for excellent audio quality

VOICE TECHNOLOGIES

CPSIA information can be obtained
at www.ICGtesting.com
Printed in the USA
BVHW022046260822
645596BV00001B/8